Puppy Training Step-By-Step

3 BOOKS IN 1

Puppy Training, E-collar Training And All You Need To Know About How To Train Dogs

PAUL DAVIS

© Copyright 2019 - All rights reserved.

The content contained within this book may not be reproduced, duplicated, or transmitted without direct written permission from the author or the publisher.

Under no circumstances will any blame or legal responsibility be held against the publisher or author for any damages, reparation, or monetary loss due to the information contained within this book, either directly or indirectly.

Legal Notice:

This book is copyright protected. It is only for personal use. You cannot amend, distribute, sell, use, quote, or paraphrase any part or the content within this book, without the consent of the author or publisher.

Disclaimer Notice:

Please note the information contained within this document is for educational and entertainment purposes only. All effort has been executed to present accurate, up to date, reliable, complete information. No warranties of any kind are declared or implied. Readers acknowledge that the author is not engaging in the rendering of legal, financial, medical, or professional advice. The content within this book has been derived from various sources. Please consult a licensed professional before attempting any techniques outlined in this book.

By reading this document, the reader agrees that under no circumstances is the author responsible for any losses, direct or indirect, that are incurred as a result of the use of information contained within this document, including, but not limited to, errors, omissions, or inaccuracies.

Table of Contents

BOOK 1: THE COMPLETE GUIDE TO DOG TRAINING 11

Introduction: Training Your Dog.. 13

Chapter 1: Why Should You Train Your Dog? 15
- Important Reasons Why Dog Training is Essential............................... 16
- What Happens When You Don't Train Your Dog?............................... 19
- FAQs About Dog Training 22
- Dog Training Mistakes to Avoid 26

Chapter 2: Dog Training Basics 31
- Different Methods of Dog Training 32
- Tips for Getting Started 36
- Things You Need for Dog Training.. 40
- Basic Commands and How to Teach Them 43

Chapter 3: Training Puppies 50
- How Early Should You Start? 52
- Coming Up with a Training Schedule for Your Puppy 53
- Potty-Training Your Puppy 60
- Practical Tips for Training Puppies 63

Chapter 4: Training Young Dogs 69
- Learning from Older Dogs 70
- Top Tips for Training Young Dogs 71

Chapter 5: Training Adult Dogs 76

    House-Training Adult Dogs 81

    Basic Obedience Training for Adult Dogs 83

    Adult Dog Training Tips from Experts 84

    Training Adult Dogs to Do New Tricks 86

Chapter 6: Training Senior Dogs **87**

    Should You Train Senior Dogs? 88

    Things to Remember When Training Senior Dogs 89

    Helpful Tips for Training Senior Dogs 93

Chapter 7: Training Different Dog Breeds **97**

    Dog Breeds and Behaviors 97

    The Easiest Dog Breeds to Train 99

    The Most Challenging Dog Breeds to Train 103

    General Tips for Training Different Dog Breeds 107

Chapter 8: Common Behavioral Issues in Dogs and How to Deal with Them **111**

    Training Hyperactive Dogs to Calm Down 111

    Play Biting, Mouthing, and Nipping 113

    Dealing with Aggression 116

    Chewing, Digging, and Other Destructive Behaviors 118

    Barking, Whining, and Other Noisy Behaviors 120

Chapter 9: Rewards and Punishments **122**

    Reward-Based Training 124

    Positive Reinforcement Training 126

    Should Punishments Be Part of Your Training? 128

    Effective Ways to Discipline Your Dog 130

Chapter 10: Advanced Training Tips and Tricks **133**

Precision Dog Training 134
Tips for Training Working Dogs 136
Training Small Dogs 137
Training Large Dogs 138
More Advanced Training Tips and Tricks 140
Conclusion: Training Your Dog the Right Way 142
References 144

BOOK 2: E-COLLAR TRAINING STEP-BY-STEP 157
Introduction 159
Chapter 1: Choosing the Right Dog for You 162
Tips for Choosing the Best Dog for You 162
Choosing the Best Breed for Training 166
Best Breeds for Training 166
Where to Go? 172
Tips to Keep in Mind Before Bringing Your Dog Home 174
Chapter 2: E-Collar Basics 178
What Is the E-Collar? 178
Types of E-Collars 179
E-Collar Accessories 182
How to Use an E-Collar 183
E-Collar Safety 187
Benefits of an E-Collar 188
Common Myths About the E-Collar 190
Chapter 3: Your Dog and Their E-Collar 192
Choosing the Best E-Collar 192

The Fundamental Five 192

Your Dog's Reaction to the E-Collar 194

Chapter 4: What You Need to Know Before Training Begins 201

General Training Tips for Dogs at Any Age 201

Training Your Puppy.. 212

Training Your Older Dog... 215

Chapter 5: Let the Training Begin 218

Tips to Prepare Your Dog for Training 218

Basic Commands... 219

Home Base and Perimeter Training.................................... 221

Sit Training... 226

Lay Down Training... 227

Come Training.. 229

Stay Training... 231

Get Down Training... 232

Chapter 6: Training Strategies, Levels, and Your Dog 234

Training Levels... 234

Practice Real Life Training 237

Intermediate Level Tricks for Dogs 245

Spin and Twist.. 249

Chapter 7: Advanced Training with E-Collars 252

Will I Stop Training? 252

Readjusting the E-Collar for a Growing Dog...................... 253

Agility Training.. 253

Basic Training Before the E-Collar 254

Boing... 257

Hunting 258

Chapter 8: Common Mistakes **260**

    Lack of Consistency in Training 260

    Training Your Dog for Too Long 260

    Dog Owners Don't Send a Signal Immediately 261

    People Become Codependent on the E-Collar 262

    The Dog Has Not Received Any Type of Prior Training 262

    People Don't Understand How the E-Collar Works 263

Chapter 9: Frequently Asked Questions and Answers **265**

    Question #1: Do Different Breeds of Dogs Learn Differently? 265

    Questions #2: Do I Wean My Dog Off the E-Collar? 265

    Question #3: How Do I Know the E-Collar Isn't Harming My Dog? 266

    Question #4: How Do I Know When to Start Using the E-collar? . 266

    Question #5: I Have Seen Other Dogs React Negatively to the Shock from the E-Collar. How Do I Know My Dog Won't? 267

Conclusion **268**

References **273**

BOOK 3: TRAINING YOUR PUPPY STEP-BY-STEP **277**

Introduction **279**

Chapter 1: Choosing the Right Dog **281**

    Picking Where Your Dog Comes From 282

    Energy Levels of Certain Dogs 286

    Child-Friendly or Not 291

    Interacting with Other Animals 293

Grooming Needs 298

Other Breed Specific Conditions to Remember 300

Chapter 2: Preparing Your Home for Their Arrival **303**

Starting Positive Habits from the Beginning 303

The Right House Setting 306

Tools Needed 308

The Training Room for Your Dog 311

Hiding Certain Belongings 316

Choosing the Right Toys 318

Chapter 3: Principles to Train Dogs **321**

Potty Training 321

Basic Training Rules 326

Positive Reinforcement and Remaining Patient 329

Specific Rules for Small Dogs 333

Specific Rules for Medium Dogs 334

Specific Rules for Large Dogs 334

Chapter 4: Training Your Puppy Outside of Your Home **336**

Going for Walks and Leash Training 336

Doggy Daycare 340

Parks 342

Friends' Homes and New Environments 345

What to Do When You are Not Home 347

Public Tips and Reminders 349

Chapter 5: Skills that Dogs Need to Know **351**

Sitting and Laying Down 351

Bark Control 354

- Greeting Visitors and Walking Properly.. 356
- Knowing Their Name and Coming on Command 358
- Fun Tricks for Your Dog.. 360

Chapter 6: Kinds of Exercises for Your Puppy **363**
- Training and Agility Classes 363
- Swimming 365
- Catch 367
- Dog-Led Walks 369
- Creating a Routine 369

Chapter 7: What to Do for the Health of Your Dog **372**
- Spay and Neutering 375
- Picking the Right Food and Treats 377
- Avoiding People Food 380
- Flea Treatments and Bathing 381

Chapter 8: What Not to Do When Training Dogs **385**
- Aggression in Training 385
- Inconsistencies 388
- Useless Repetition 389
- Confusing the Dog 390
- Abandoning a Ritual or Habit 392
- Treating All Dogs, the Same 393
- Positive Reinforcement with Bad Behavior 394
- Not Proofing Tricks 395
- Waiting Too Long 397

Chapter 9: Your Step-by-Step Training Plan **398**
- Get to Know Your Dog 398

Create a Schedule	398
Keep it Simple	400
Work on Rewards	401
Repeat, Repeat, Repeat	401
Check-In with Your Dog	401
Conclusion	**403**
References	**406**

BOOK 1

THE COMPLETE GUIDE TO DOG TRAINING

A How-To Set of Techniques and Exercises for Dogs of Any Species and Ages

PAUL DAVIS

Book 1 - The Complete Guide To Dog Training

Introduction: Training Your Dog

Fig. 1: Trained Dog. From Unsplash, by Reed Shepherd, 2018, https://unsplash.com/photos/r1q76Rut5t8 / Copyright 2018 by Reed Shepherd /Unsplash.

Whether you are a new dog owner or are just thinking about bringing a furry friend into your home, you should know that training your dog is an essential part of the process. Dog training is more than simply teaching your dog commands and tricks. It involves intentional communication with your canine companion. Training your dog requires a lot of practice, which means that you will have to communicate continuously for life! Although socialization with other dogs helps with domestication, your dog should also learn how to properly communicate with people. The most effective way to do this is through training—a clear and consistent form of canine education with a foundation of mutual respect and trust.

There are different ways to train dogs, and it's up to you to find the most effective method for you and for your furry friend. To do this, you should try to achieve a balance between yourself and your dog in their life and in their environment. To achieve balance, you should learn how to communicate effectively with your dog; create structure in your dog's environment and in your relationship with them, improve the relationship you have established, and reduce the natural latent stress that dogs experience in their lives.

In doing all of this as part of your training, your pet will start paying more attention to you. In fact, your dog may even start looking forward to obeying you or following your lead. Training your dog won't be a short or easy process. You must invest a lot of time and money into it. Of course, this doesn't mean that you can't have fun along the way too! In this book, you will learn everything you need to know about training your dog. From the basics to more advanced tips and everything in between, our discussions will help make dog training much easier for you. Whether you have a puppy, a young dog, an adult dog, or even an older one, we've got you covered.

The more time you spend training your canine companion, the more you will see your bond strengthen. Focusing intently on your dog while training makes your dog trust you more profoundly. Over time, you may even learn how to anticipate your dog's behavior, thus making it easier for you to train them. Furthermore, having a well-trained dog means that you can allow them to explore the world freely without having to worry about their safety—or the safety of the people around them!

Book 1 - The Complete Guide To Dog Training

Chapter 1: Why Should You Train Your Dog?

Dog training is an important aspect of a dog's life, and as a pet owner, it's your responsibility to ensure that it is done properly. Training provides your dog with mental and physical stimulation to keep them happy, healthy, and well-rounded. Positive training techniques are enjoyable for dogs, and they help strengthen your relationship. For such techniques, you will set-up your dog to succeed, then reward them for their good behaviors. The rewards you give your canine while training may come in the form of treats, verbal praise, or even physical affection. Each time your dog does something good, especially after you have taught the action, you can give the reward.

In general, many dog trainers who use these kinds of training methods also ignore "bad" behaviors. That way, your dog won't be rewarded in any kind of way—sometimes, paying attention based on bad behaviors seems like a "reward," which is why ignoring them is much more effective. Therefore, if your dog doesn't receive a reward for doing something (because it's bad) and they don't receive attention either, the chances are that they will stop doing the behavior. For instance, if you have an overly excited dog who loves jumping up when you get home, you can ignore your dog throughout this specific behavior. Once your dog gets back down on the floor, that's when you look at them and give them positive attention. In instances where you come home, and they greet you without jumping, then you can reward them for this.

One thing you must avoid when training your dog is using physical punishment. Contrary to some beliefs, hitting your dog doesn't actually stop bad behaviors. Most of the time, physical punishments will even make matters worse!

Seeing as dog training is the responsibility of owners, learning everything you can about how to train your dog is crucial. And the earlier you start, the more beneficial for yourself and your dog.

Important Reasons Why Dog Training is Essential

Through the years, dogs have become known as man's best friend, and the main reason for this is that they are social and pack animals. Dogs kept as pets look to their owners to guide them so that they can learn how to behave as expected. Unless you teach the rules to your dog, don't expect them to know what your rules are. Dog training is essential as it provides your canine with the education they need to be a well-behaved, functional member of your family. Here are some of the most important benefits of dog training:

- **Your dog will learn how to listen to you**

  While your dog may listen to you when you call their name, this doesn't mean that they will follow your orders. But through training, you'll discover that your dog is listening to you—really listening. Although they won't listen every time you say something, your dog will learn what the commands you say really mean. Then, they will start following you more frequently than before.

- **You will be able to control your dog more easily**

  Later on, you will learn about the basic commands to teach your dog—and how to teach them. Training your dog to learn these basic commands will help you control your dog easily and more effectively, no matter what the situation is. Whether you are at home or decide to take your dog out for a walk, these commands will also keep your canine companion safe.

Book 1 - The Complete Guide To Dog Training

- **Dog training can save your dog's life**

    When dogs get scared, they tend to run away, and this can be extremely dangerous for them, especially if they run into the streets. But when you have trained your dog to come back when you call their name, this can save their life.

- **It gives you a lot of opportunities to have fun with your dog**

    While you are training or during breaks, you can have a lot of fun with your dog. For training, you can find fun ways to teach your dog basic (or even advanced) tips and tricks. During breaks, you can play with your dog. No matter what species you belong to, play is always a lot of fun!

- **Dog training provides your dog with a solid foundation**

    Teaching the basic commands to your canine provides them with a solid foundation to help you deal with different kinds of circumstances and situations. Then, as you teach your dog new tricks, you can incorporate the basics to reinforce them and strengthen your dog's learning process. Again, the longer you stick with your dog training, the more you will see improvements in how they behave and react in various situations.

- **It gives you a better understanding of your dog**

    Training your dog involves spending a lot of quality time with your them. Because of this, you will be able to gain a better understanding of your canine's unique signals and body language. And when you understand your dog better, you will be able to train them better too.

- **It strengthens your bond**

   From the get-go, you should be able to establish a strong connection with your dog; otherwise, they won't listen to you. After establishing that strong connection, training your dog using the proper techniques and methods will help make your bond even stronger. The great news is that once you have trained your dog well, you will notice that they are more confident, relaxed, manageable, and content. Your dog will also trust you more, thus making them happier and more willing to listen.

- **Training your dog with other dogs teaches valuable socialization skills**

   You don't have to train your dog alone. Once in a while, you can bring your dog to the park and join other pet owners who are training their dogs too. You also have the option to enroll your pet in obedience classes to learn with other dogs. Either way, when your dog is trained with other dogs, they learn how to socialize with each other—another essential skill for a well-rounded canine.

- **Dog training provides the stimulation needed to stay happy and healthy**

   Training provides your canine with many different types of stimulation. These include physical exercise, mental stimulation, quality time with you, and you may even teach them responsibility. Apart from these, your dog will also enjoy all of the rewards they get each time they do something right.

As you can see, dog training is beneficial for you and for your dog as well. With all of these benefits and more, you should see why it's important to train your dog as soon as possible. You can start enjoying all of these benefits while helping your dog grow into a well-rounded, well-mannered, and happy adult dog.

What Happens When You Don't Train Your Dog?

Fig. 2: Growling Dog. From Pixabay, by Rain Carnation, 2015, https://pixabay.com/photos/dog-learn-tricks-training-hybrid-622705/ / Copyright 2015 by Rain Carnation /Pixabay.

No matter how much you love your dog, there will be things they do that you will not approve of, and if you want things to change, training is key. An untrained dog may think that they are in charge at home, and they might end up driving you crazy! They bark incessantly, chew up all of your shoes, and run after your visitors nipping at their heels. If you're unwilling to train your dog, you might just throw them a toy or some other objects just to get them to stop.

Does this situation sound familiar to you?

Well, if you continue doing this instead of putting in the effort to train your dog, don't expect things to improve. When dogs act inappropriately or do "bad" things, the problem doesn't lie with them; it lies with you, their owner. For untrained dogs, they do these things because it's part of their nature. Unless you teach your dog the "right" things, the behaviors you dislike will continue. And this might make living with your dog extremely challenging.

Just as you wanted to make your dog a member of your family, your dog wants you to step up and lead. The truth is that dogs have an innate need for security and social structure. Unless you take the role of the leader, your dog will feel like they are in charge, and this is when the bad behaviors may start. Here are some things that may happen when you don't train your dog:

- **Your dog gets bored**

    Bored dogs engage in activities such as barking, digging, chewing, pacing, and jumping on you. This is why one of the most important benefits of dog training is stimulation. Dogs need to be stimulated so that they won't act out. A bored dog has nothing better to do, so they will simply do whatever they feel like!

- **Your small dog will develop "small dog syndrome"**

    Small dogs don't know that they are small. Just because you have an adorable toy dog, that doesn't mean you don't have to train them. Without the proper training, your canine companion might turn into a nasty, short-tempered terror!

- **Your big dog will become a wrecking ball**

    While small dogs may become prima donnas, big dogs who haven't been trained can start destroying everything around them! If you have a big dog at home, don't let their size intimidate you. Training is still important no matter how big your dog is, especially if you want to keep your home and belongings intact.

- Your dog will start landscaping your yard

    All dogs love to dig as part of their exploration. They dig to investigate things they smell, they dig when searching for food, and they may even dig to escape your backyard. Digging is a destructive and dangerous behavior that won't go away unless you train your dog that this isn't okay.

- **Your dog will chew everything in sight**

  Dogs love chewing things, especially while they are puppies. They don't do this to annoy you; they just like to chew! Unfortunately, unless you give your dog a chew toy or you train your dog not to chew your things, you might come home to see your pillows, the legs of your chairs, and the cords of your electronic appliances all chewed up!

- **Your dog will keep you up all night by howling and barking**

  It's good if your dog barks or howls when they spot an intruder in the vicinity. But when they do it all day and all night, it will surely drive you mad. The good news is that you can actually teach your dog to bark only when appropriate. How do you do this? Through dog training, of course!

- **Your dog will have accidents all around your house**

  Although you may have already housebroken your dog, don't think that this is all the training they need. Otherwise, you might notice that your canine companion is leaving you surprises all around your house. House-training is one of the most basic—and most important—things to train your dog. And you need to keep practicing this so that your dog doesn't forget. Also, you should continue training and stimulating your canine even after they have learned a lot of tricks, as this provides them with the stimulation they need.

- **Your dog becomes sad**

    Just like us, dogs can get the blues, especially when they experience a major loss or change in their lives. To perk up your dog, start a training schedule. This gives you time to communicate and interact with your dog, which are activities they really enjoy. If you notice that your dog is feeling sad, play with them, train them, and talk to them to lift their spirits. But if the blues go on for more than a few days or weeks, you should consider taking them to the vet.

As you can see, not training your dog can lead to many negative consequences. This is why dog training is considered the responsibility of the owners. Since dogs cannot train themselves, you have to step up to the challenge!

FAQs About Dog Training

Before you learn the basics of dog training, you may have some questions in your mind about the whole process. As previously mentioned, dog training is an essential part of a dog's life; therefore, as an owner, you must make this happen. When you bring a dog into your home and make that dog a part of your life, you must also teach them how to fit in. Dog training also helps you learn about your dog's needs and personality. The more you learn, the more you can make adjustments in your expectations to strengthen your bond and improve your relationship.

Take the time and put in the effort to train your dog to avoid frustrations and ensure a happy life with your canine companion. To help you understand dog training even further, let's answer some of the most common questions people ask about it.

1. **What's the best method for training dogs?**

While there are many methods you can choose from when it comes to dog training, make sure to focus on those that are dog-friendly. This type of training is all about learning how to understand your dog and meet all of their needs. Remember that dogs are individuals, just like us. This means that although there are general methods for training, you must make sure that you customize your methods to make them kind, gentle, effective, and humane. Dog-friendly training methods encourage positive behaviors while ignoring the negative ones. These methods don't incorporate coercion, intimidation, or physical punishment that causes emotional, mental, or physical pain to your furry friend. Even if you think that your dog has "behavioral problems," it's never a good idea to use such methods, as these can make the situation worse. Beyond this, it is up to you as an owner to determine the best method for training your own dog. You may have to do some experimentation to see what works and what doesn't.

2. **How to train dogs in a dog-friendly way?**

The first thing you must do is to learn everything you need in terms of the methodology and principles of how to teach the behaviors you want your dog to have. Fortunately, you have purchased this book that contains a wealth of information to help you learn about dog-friendly training. You can also watch videos online or attend dog training classes to reinforce everything that you learn here.

3. **How to stop your dog from jumping up at you each time you get home?**

When your dog jumps up at you when you get home, this shows how excited they are that you have finally returned. Unfortunately, this isn't an ideal behavior. As part of your dog's training, you train your dog to jump on you only when you give a specific command. For instance, when you say something like "Play," then your dog can jump up. But once you give another command such as "Stop" or "Enough," your dog must stop. Practice these commands with your dog to make them stick.

4. **How to stop your puppy from whining or crying at night?**

This is an easy one. To stop the whining, crying or barking, you simply have to ignore it. Although this can be extremely challenging, it is the most effective thing to do. If you keep checking on your puppy throughout the night, shush them, or do anything else to acknowledge their whining, they will keep doing it. But when you ignore these behaviors, over time, your puppy will learn that whining, barking, or crying won't get them anywhere.

5. **How to correct your puppy or young dog when they have a house-training accident?**

The only time you should correct your puppy or young dog is when you catch them in the act. Young dogs learn by association. This means that when you see your young dog or puppy pooping or peeing where they shouldn't be, correct the behavior right away. When you see the "surprise" left by your young dog but they are nowhere in sight, there is no point in trying to correct them. While your puppy is performing the unwanted behavior, clap your hands while saying "NO!" in an urgent tone. Then carry your puppy to where they should be doing their business and give the proper command.

6. **Why do some people opt for dog training classes?**

Some people might not have the time, patience, or knowledge to train their dogs on their own. This is probably why they enroll their pets in dog training or obedience classes. While these programs are very effective, training your dog at home comes with more benefits. For one, you will be able to learn more about your dog. You will also have the chance to spend quality time with your dog each day, thus strengthening your bond. There is also the benefit of knowing exactly what your dog is capable of and what you need to improve on. Also, these dog training classes may cost a lot of money! One great benefit of these classes is that your dog will have a chance to socialize with other dogs while training. But you can have the same benefit when you train in the park or when you train with your friends who are training their dogs too.

7. **Should you wait until your puppy has received all of the required vaccines before you start training?**

This is a very common misconception that causes dog owners to delay their dog training. Yes, it's important to make sure that your puppy gets all the necessary vaccines. But you can start training your puppy even before the completion of these shots. Just make sure that you continue taking your puppy to the vet throughout your training until your little friend has received all the vaccinations they need.

8. **What is clicker training?**

This is a type of motivational, positive training method that has its basis in operant conditioning and the scientific principles of learning theory. The fundamental principle of this training method is that rewarding behaviors reinforce or "make stronger," therefore, encouraging the subject to do them more often. For clicker training, you would use a tiny mechanical clicker to indicate the precise moment when your dog is doing the behavior you requested. Then. you would immediately give a reward after the sound. The great news about clicker training is that it works effectively for all dog breeds!

Dog Training Mistakes to Avoid

Dog training doesn't just provide essential stimulation for your dog; it also teaches them how to behave well. But when it comes to training dogs, there are some mistakes dog owners make that may cause more harm than good. For one, over-training your canine is never a good idea. This might undo all of the training your dog has already learned along with the time and effort you have already invested. To give you a better idea of the mistakes to avoid, here are the most common ones:

- Not considering your dog's age

    While training your puppy while still young is a good thing—especially since early life experiences are crucial—you should never set intense and long training sessions for your puppy. When you start off with difficult training sessions or you try to teach too many things, this will have adverse consequences on your young dog. While they may obey you, they do so fearfully and grudgingly. Then, they grow up that way instead of being eager, enthusiastic, and happy about pleasing you.

    Consider your training carefully if you have a senior dog too. While it's possible to train older dogs, they may suffer from medical conditions and muscle or joint soreness that might make learning difficult for them. Some senior dogs also experience a decrease in their cognitive abilities as they age. Therefore, when training a senior dog, you may need more patience and time before you will see results.

- Not following the training schedule you set for your dog

    In the beginning, you may feel very enthusiastic while training your dog. You set a schedule that includes short, frequent sessions and follow it religiously. Over time, you may get tired of your training sessions, especially when you see that your dog has already learned a few tricks. Because of this, you start missing your training sessions until you forget them altogether.

Unfortunately, when you don't train your dog frequently, this causes them to go on "auto-pilot." Over time, your dog might forget the things they've learned, and the behaviors you have worked to eliminate have come back. If you want your training to stick, and you want your dog to have the stimulation they need, make sure that you train regularly and consistently.

- Repeating commands until your dog obeys

  This is a very common mistake. You try to teach a command, such as "Stay," but your dog is distracted or confused, so they don't respond. So, you keep repeating the command until your dog performs the action halfheartedly. What your dog learns here is to keep stalling until you have repeated the command several times—and this learned behavior is very hard to unlearn.

  Say a command once. If your dog ignores you, they might not have heard you, they may be distracted, you might not have taught it correctly, or your dog is simply rebelling against you. Bring your dog to a quieter environment and say the command again. If you are still ignored, re-teach the command until your dog learns how to do it again. But if your dog follows you, praise them!

- Generalizing your training instead of customizing it for your own dog

  Each dog has a unique behavioral profile and personality. Instead of generalizing your dog, try to understand them more before you proceed with your training. Experiment with different training methods until you determine which method is most effective with your canine companion. If you discover that some methods don't work, make the necessary adjustments. If you see that your dog enjoys a particular method of training, find ways to incorporate this method while teaching other behaviors, lessons, or tricks to your dog. Even though there are general or basic methods for training, learning which ones work for your dog is key to the success of your dog training.

- Trying to do too many things, too fast

Have you ever felt overwhelmed at work when you were assigned too many different tasks to complete by the end of the day? Well, dogs can feel overwhelmed too! Some owners try to teach too many things at once or force their dogs to

Fig. 3: Overwhelmed. From Pixabay, by Ella_87, 2019, https://pixabay.com/photos/dog-cavalier-king-charles-animal-4246782/ Copyright 2019 by Ella_87 /Pixabay.

train for hours on end just to speed things along. However, this isn't good for your dog. In fact, when your dog gets overwhelmed, they won't be able to learn anything. Try focusing on a maximum of two commands or tasks for each training session. Keep practicing until your dog has actually learned these commands before moving on to new ones. Remember, patience is key.

- Giving too many treats and not giving praise

While dogs love treats, and giving them reinforces good behaviors, giving too many won't be good for your dog. This might cause a fixation on food instead of learning the behaviors themselves. Try to balance the rewards you give. Give your dog treats, praises, and affectionate gestures as needed. Try to determine which is appropriate to get the best response from your dog at each training session.

- Not having enough confidence while training

When you're not confident enough, your dog will see this as a weakness. They sense this instinctively, and when your dog senses this lack of confidence in you, they might not listen to or follow you. Your dog might even exploit this weakness of yours because it is their nature. To increase your confidence, you may want to attend dog training classes, especially at the beginning of your training. When you feel more confident to train on your own, start at home. The more confidence you gain, the more you can try training your dog in different environments.

- Being too emotional

Finally, when you show too much emotion while training, this tends to have an effect on your dog's learning ability. For instance, when you are angry, irritated, or forceful while training, you might intimidate your dog, making them fearfully obedient. Conversely, when you show too much energy, squeal with delight each time your dog does something right, or jump around as much as your dog, this will have an adverse effect on how they learn and focus. While training, it's best to show calm indifference. Have a demeanor that suggests easy authority and competence. This laid-back kind of energy where you show love and guidance calms your dog down and helps with your confidence too.

Chapter 2: Dog Training Basics

Once you bring a dog into your life, you won't be able to imagine your life without them. Dog owners adore and admire their canine companions for their enthusiasm, unconditional affection, and loyalty. But when man's best friend keeps chewing things up, barking all night, digging the yard, and so on, you might not appreciate them as much. If you want to make the most out of your bond with your canine, you must train them to help you live harmoniously. When you train your dog, it improves both your lives, and it also ensures both your safety. The great thing about dogs is that they love to learn—and you can teach them through good communication. You should help your dog understand how they should behave and why following you is in their best interest.

Any responsible dog owner should learn how to train their dog. Training is also a fun and enriching way to bond with your canine companion. Part of your dog's training is to set clear boundaries. While training your dog, make sure that you clearly define these boundaries so that your dog learns to respect them. To do this, you will have to train your dog to follow commands that are very specific. This makes it easier for you to enforce the boundaries you've set consistently. Just like any other skill, you must start with the basics when you're learning how to train your dog. The same thing goes for when you're actually training—start with the basics, practice, then move on when you think you (or your dog) are ready for the more advanced stuff. Looking at dog training as an adventure to take with your canine companion makes it more fun for both of you. And the more you learn before you start training, the more confident you will be throughout the process.

Different Methods of Dog Training

When it comes to dog training, there are many different methods you can choose from. But when you know nothing about these methods, how can you determine which one to use for your dog? Hearing about all of these methods might make you feel confused and overwhelmed. Don't worry, because all dog owners experience this, especially at the beginning of their dog-training journey. For you to choose the best method for training your dog, you'll first need to learn more about them. Here are the most common training methods for you to consider:

- **Alpha Dog Training (Dominance Training)**

  For this type of training, you would depend on the natural pack mentality of your dog. Since this is their instinct, you can use it to build a relationship of appropriate dominance and submission. This training method is based on the theory that dogs consider you and your whole family as a pack—their pack. Here, you must establish your role as the alpha in order for your dog to respect and obey you. If you plan to use this method, you will have to understand the body language of your dog and learn how to respond appropriately while projecting confidence and authority. Since there is a struggle of dominance here, you need to put emphasis on reinforcing your dog's training consistently.

- **Behavioral Training**

  This method of training refers to any kind of training method where dogs learn how to be well-behaved around people and around other animals. Behavioral training may include teaching basic commands too, but your main goal here is to teach your dog how to be a "good citizen." Some trainers use this method to reduce or eliminate behavioral issues. This is the most common and most basic training method for puppies and young dogs.

- **Positive Reinforcement Training**

This training method was popularized by more modern dog trainers. It's a fairly straightforward practice for training dogs as it involves rewarding good behaviors. Here, you would encourage your dog to repeat good behaviors by immediately giving rewards after they are done. When your dog does something bad, you would either ignore the behavior or take away their rewards.

- **Obedience Training**

This type of training method focuses on teaching your dog how to obey you by following the commands that you use. It's a bit more advanced than behavioral training, but it can help you deal with behavioral problems even before they begin. This is another training method that's ideal for young dogs.

- **Scientific Training**

Coming up with a definition for this training method is quite challenging since it relies heavily on information that is constantly changing and building. Your aim here would be to understand the nature of your dog, their abilities, and how effective punishments and rewards would be when used. The information for this training method comes from animal experts and behaviorists who continuously create new experiments and studies to help us understand the psychology of our four-legged friends. Before you can correct your dog's behaviors, you must understand them first.

- **Agility Training**

This is the perfect training method to use if you plan to let your dog participate in different dog sports, such as jumping, racing, and maneuvering through obstacle courses. This training method is more advanced, so your dog should already know and understand the basics before you begin. During competitions, you won't be allowed to reward or touch your dog. This means that you should have a very strong bond with your dog so that they can complete the sports by taking cues from your physical gestures and verbal commands only.

- **Clicker Training**

This method of training has its basis on operant conditioning, much like positive reinforcement training. As a matter of fact, some trainers believe that this training method may be part of the positive reinforcement training method. The only difference is that you would use a small mechanical device that makes quick and sharp noises to signal to your dog that they have performed the behavior you have requested.

- **Vocational Training**

Just like us, dogs have the capacity to learn different skills. Some dogs learn how to hunt, do rescue work, assist handicapped individuals, and more. If you want your dog to have such skills, you can use the vocational training method. For this method, you will teach your dog very specific commands and techniques to communicate with the people around them and enhance their senses. You should know, however, that vocational training is both time-consuming and rigorous.

- **House-Training (with or without Crate-Training)**

This training method is essential, especially if you want your dog to live indoors. Basically, you would train your dog to "do their business" in the proper locations. This is one of the first methods of training you must do. While you can do house-training on its own, crate-training can make the process easier. This is where you would keep your dog contained (inside a crate) to strengthen your training and make it flow smoothly.

- **Model-Rival Training (Mirror Training)**

This method of dog training relies on learning through observation. You can do this by allowing your puppy or young dog to observe an adult or older dog who has already been trained. This provides a good role model for your dog to mimic. If you don't have another dog, you can also ask one of your friends or family members to act as the role model and pretend as if you are training them. As your dog observes, they learn the correct behaviors from the model.

- **Leash Training**

All dogs must learn how to walk on a leash—unless you plan to keep yours confined within your property, which, incidentally, isn't good for the dog. When you take your dog outside, you must use a leash to comply with the local laws in most areas and to keep your dog safe as well. This is another essential training method that teaches your dog the proper way to walk while on a leash. This makes walking a more enjoyable experience for yourself and your dog too.

- **Relationship-Based Training**

This method of training is a combination of other training methods. For this, the focus is on an individualized approach for yourself and your dog. The driving force in this training method is the relationship that exists between you and your canine.

Therefore, the stronger your relationship is, the more effective your training will be.

While there are other types of dog training methods out there, these are the most common—and the ones that don't use physical or harsh punishments. Now that you know the basic information about these methods, you can have a better idea of which training method you want to use or which one would be most suitable for yourself and your canine companion.

Tips for Getting Started

Fig. 4: Getting Started. From Unsplash, by Wyatt Ryan, 2017, https://unsplash.com/photos/fkLr2QOQitk / Copyright 2017 by Wyatt Ryan /Unsplash.

So, you finally take your dog home. After a few days, it's time to start your dog's training. Ideally, you would have already learned all that you can about dog training by attending dog training classes, watching dog training videos, and reading dog training literature while you were still thinking about bringing a dog home. Of course, if you're reading this book and your dog is already in your home, that's okay too. The important thing is to educate yourself first before you can start educating your furry friend.

If you bring home a puppy, you must understand that puppies aren't small adult dogs; they are infant dogs. Therefore, you shouldn't expect too much from your puppy. Instead, consider their mental and physical limitations. When it comes to choosing the name of your dog, do so wisely. Choose a respectable name that you won't feel ashamed of even when you're training your dog in public places. Also, choose a name that makes you feel good, not one that other people might laugh at.

As soon as you bring your puppy home, give them a bottle filled with warm water along with a ticking clock. Place these items in your puppy's sleeping area as they will imitate the heartbeat and heat of your puppy's littermates. This will help soothe your puppy, especially at the beginning. Even if you bring home a young, adult, or senior dog from the shelter, there are many things you can do to make your new dog comfortable in their new home. After this, it's time to start preparing for your dog training sessions. For this task, here are some pointers to start off with:

- **Prepare a pen, baby gates, or a crate for your puppy**

    Whenever you aren't around to supervise your puppy directly, you must place them in a safe place, such as a pen, a crate, or a room closed off with baby gates. Then, provide safe toys for your puppy to play with or chew on while you're not around. Make sure that the area you place your puppy in doesn't contain any dangerous items or objects that your puppy might end up destroying. This helps ensure that your puppy won't learn any bad habits early on.

    Also, think about giving your puppy their own room. This should be a place where your puppy can sleep and have "alone time," a place that isn't being used by any other person or pet. If your puppy stays in their "room" quietly, you can reward them for it. You can also use a crate as their room. Crates are also very helpful when it's time for you to start house-training.

- **Remember that dogs don't understand English—or any other language**

    Puppies aren't born with the innate ability to understand our language. So, when you bring your puppy home, don't expect them to understand you when you say "no." If you want your puppy to understand what this word means, demonstrate what you want them to do when you say "no." Also, dogs don't think like us humans. They don't do things to anger you, they don't make plans for revenge, and they don't hold grudges. They simply do what makes them feel safe or happy; this is their nature.

- **Provide your dog with enough mental and physical stimulation**

When dogs feel bored, they tend to get into trouble—no matter what age they are. This is why it's important for you to provide enough stimulation. The best way to do this is through short but meaningful training sessions. In particular, mental and physical stimulation helps with the growth and development of puppies.

- **Try catching your dog doing good things**

It's important to observe your dog as much as you can to get a better sense of their personality and behaviors. While it's a lot easier to catch dogs doing bad things than scolding them after, how often do you catch your dog doing something good? You should be able to catch these instances as well because this is the perfect time to show your dog that what they just did is the right thing.

After seeing behavior that you approve of, it's time to reward your dog using positive reinforcement. This can come in the form of treats, toys, affection, praise, and other good things. Rewarding your dog sends them a message that they are doing the right things. Just make sure that the rewards you give come immediately after the good behavior. Otherwise, you might be rewarding a different behavior altogether.

- **Once in a while, use high-value treats while training**

  Dog treats will always make your dog happy. But once in a while, try giving your dog high-value, healthy treats like liver, cheese, or chicken breast. You might be surprised at how hard your dog will work to get such treats. While store-bought dog treats will be effective while training in environments free of distractions, when you plan to train in new environments, you may want to bring better treats to keep your dog engaged. Just make sure the treats are soft and easy to chew so that you don't have to take long pauses while training.

- **Set "dog time" for your training sessions when your sole focus will be on your dog**

  One of the best things about dogs is that they live in the moment. The most important thing to them is what is happening right now. A few minutes later, that important moment is already forgotten. This is why you should focus on your dog during your training sessions. Think of this as "dog time," when you shouldn't be doing or thinking about anything else.

- **Always end your training sessions positively**

  This is an excellent way to keep your dog excited about your training sessions. Throughout the training session, your dog tried their best to impress you and make you happy. So, you should end your session with a few minutes of play, some petting, a treat, or plenty of praise.

- **Show happiness whenever your dog approaches you**

  Whether you are training, called your dog over, or didn't call your dog, when your dog comes to you, show them how happy you are. Call your dog's name in a playful and happy tone, and if they came to when you called, don't forget the reward!

Things You Need for Dog Training

After deciding what type of method to use and preparing your home, your dog, and yourself for training, it's time to gather all the basic tools you need for the task. The great thing about these basic tools is that you can use them throughout your training—from beginning to advanced training.

1. **Durable leash (at least 6-feet in length)**

    While you may have already bought a leash for your dog, you need one that's at least 6-feet long for training purposes. Also, choose one that allows you to get a solid grip on your furry friend. If you want a professional-grade leash, choose one that's made of sturdy leather, nylon, or cotton. Among all of these materials, leather is the most flexible and durable while being comfortable to hold too. Leather is also strong enough to withstand untrained dogs who tend to pull hard. Helping your dog to develop a positive association with their leash makes it easier for you to get them into cars or go out for walks. You can help develop this positive association by feeding your dog training treats each time you clip it on.

2. **Target stick**

    This tool is helpful when you need to teach both basic and complex tricks or behaviors. You can get a simple target stick or one with other features like a clicker that's already built-in or a target stick that collapses.

3. **Collar (or harness)**

For the purpose of positive reinforcement training or other training methods, you should consider choosing a simple collar with a flat-buckle and metal hardware. Metal buckles are more secure and durable than those made of plastic. If you have a dog with a narrow head—like a whippet or a greyhound—choose a martingale collar. You can tighten this collar enough to prevent your dog from slipping out of it but not so much that it ends up choking your dog. On the other hand, if you have a dog with a flat face—like a bulldog or a pug—you may want to opt for a harness instead of a collar.

4. **Portable mat**

    Although some don't think it's essential, this tool will provide your pooch with a safe space to relax wherever you take them. Portable mats are easy to transport and wash. You may even choose one that has a sticky bottom to provide better stability.

5. **Clicker**

    If you plan to use clicker training, you definitely need to get a clicker for yourself. Use this to indicate when your dog has performed the good behavior you have asked for. While clicker training isn't the only method you can use, it is a very effective one. You can even combine the use of a clicker with other positive training methods to enhance them.

6. **Long line**

    If you want to start practicing off-leash behaviors, but you're not confident enough that your dog will follow you, then you may use a long line. This is a simple and safe alternative you can use before completely taking the leash off during training. Also, you can use a long line to allow your dog to explore without you or allow them extra room for scent detection and other activities. Most long lines vary between 15 and 30 feet in length.

7. **Training treats**

    Training isn't fun without treats! In some cases, using treats makes training more effective too. There are many different types of dog treats available for dogs, but as mentioned in the previous section, you can also opt for high-value treats to make your dog feel more motivated. Vary the treats you use depending on the behavior or trick you are trying to train. Also, use different types of treats so that your dog doesn't get bored with the rewards.

8. **Training belt or treat pouch**

    While you can cram all of your dog's treats in your pocket to keep them on-hand for training, where will you put the rest of your tools and training items? Besides, you don't want all the other dogs to start following you around because you smell like treats, right? Using a training belt or treat pouch will help you out immensely as you can keep all of your items inside one. Opt for one that's odor-resistant and washable so that you can clean it as needed. There are belts or pouches that even have extra pockets for your personal items, such as your keys, phone, and wallet. You may also use a bigger container or bag to store the rest of your training equipment, including toys, treats, and comfort items.

9. **Toys**

    Apart from treats and praises, toys also serve as great rewards for dogs in training. To ensure the effectiveness of the reward, get to know your dog better. That way, you can purchase toys that you know your dog will be interested in. Keep these favorite toys on-hand during training sessions to reward your dog for a job well done.

10. Barriers

These include pet pens, pet gates, playpens, and crates. These barriers are very useful for when you want to keep your dog in a single space for chewing management, house training, and other training methods. You can also use barriers to keep your dog away from problem areas like doors, stairs, and more.

Basic Commands and How to Teach Them

The feelings of joy, love, and fulfillment that your dog will bring into your life is incomparable. Unlike most people, dogs will love you unconditionally. But without training, don't expect your dog to know exactly how to behave at home or in other environments. Training teaches your dog how to behave well while ensuring their safety. Although it's ideal for dogs to learn early on, even older dogs can be taught with the proper training. When it comes to training, there are a few basic commands your dog must learn before you can move on to the more advanced stuff. These basic commands are:

Sit

This command is one of the easiest to teach your dog, which is why most beginners start with this one.

- While your dog is standing, take a treat and hold it right next to their nose. Whenever you start teaching commands, make sure to do so in a calm manner.
- Slowly move the treat up so that your dog's head follows the treat. When this happens, your dog's bottom will lower. If needed, push down your dog's bottom for them to sit down.

Fig. 5: Dog Sitting. From Unsplash, by Spring Fed Images, 2019, https://unsplash.com/photos/t3UDG4WUxDQ / Copyright 2019 by Spring Fed Images /Unsplash.

- Once your dog is in position, give the command "Sit." Give the command firmly and clearly. Then, give your dog the treat and praise them for a job well done.

Keep doing these steps a couple of times each day until your dog can sit on command without your assistance. Then, start asking your dog to sit each time you want them to calm down—like before you leave the house, before eating meals, and more.

Stay

You can only try this command out once your dog has mastered "Sit." This is because the command starts with your dog in a sitting position.

- Start by asking your dog to "Sit." Most tricks begin with this command, which is why you should teach "Sit" first.
- Hold your palm out in front of your dog's face, then give the command "Stay." Again, say the command firmly and gently.

- Take a couple of steps back to see if your dog will, in fact, stay. If your dog stays, give them a treat as a reward.
- Each time you practice this trick, try increasing the number of steps you take away from your dog before you give a reward. Keep rewarding your pooch for staying in place no matter how brief they stay.

This is one command that exercises your dog's self-control. If your dog keeps going to you, especially at the beginning, don't feel discouraged. This is especially true for dogs who have a lot of energy and for puppies.

Come

This is an important command that helps keep your canine out of trouble. Ideally, when you give the command, this will bring your dog back to you.

- For this command, you need to attach the leash to your dog's collar to ensure that they don't run away from you.
- Go down to your dog's level and give the command "Come" as you pull on their leash gently. Strengthen the command by looking your dog in the eyes.
- If your dog comes to you, give them a treat as soon as they reach you.

Keep repeating these steps until your dog has mastered the command. Then, take the leash off and see if your dog will follow you. When training off-leash, make sure to do it in an enclosed area to keep your dog safe.

Heel

This is another important command wherein you teach your dog to walk alongside you rather than walking in front of you. While you walk with your dog, their head would be in line with your knees. This is an excellent command that teaches your dog to walk on a leash properly.

- For this command, you need to attach the leash to your dog's collar. Start off by asking your dog to "Sit."
- Hold your dog's leash with one hand and hold a squeaky toy above your dog's head with the other hand.
- While keeping the toy in position, start walking while you give the command "Heel." Don't rush this step as it might cause your dog to get excited.
- Press the squeaky toy to catch the attention of your dog while you are walking so that they start following you.
- If, at any point, your dog pulls in front of you or gets distracted, stop walking.
- When your dog's attention goes back to you, give the toy to them along with praises.
- After about half a minute of focusing on you, start walking once again.
- The more you practice, lengthen the time your dog walks with you before you give them a reward.

Practice this command daily to make it easier for you to take your dog out for walks. Over time, you may be able to walk with your dog even without a leash.

Down

As far as basic commands go, this is one of the more difficult ones to teach, mainly because it places your dog in a submissive posture. To make things easier for you and your canine companion, maintain a relaxed and positive tone while teaching this command. This is particularly important for anxious or fearful dogs from shelters.

- To start off, select a treat that has a particularly good smell, place it in your palm, and close your first around it.

Book 1 - The Complete Guide To Dog Training

- Hold your fist up to the snout of your dog. As soon as you see that your dog is interested (meaning they start sniffing your fist), slowly move your fist down, allowing your dog to follow.
- Gently slide your fist along the floor in front of your dog so that their body starts following their head. This will naturally cause your dog's body to slide down to the floor.
- Once down, give the command "Down," and give your dog a treat and some good words too.

As with the other commands on our list, it's best to practice this everyday. In cases where your dog lunges to you or tries to sit up, take away your fist and firmly say "No." You don't have to push your dog into position. Just follow these steps to encourage your dog to do it themselves.

Leave it

If you want to ensure your dog's safety, this is one of the best basic commands to teach them. Dogs are very curious, and they tend to sniff or paw at things they find interesting, even though these things may cause them harm. For this command, your goal is to let them learn that they will get a reward for leaving something (possibly dangerous) behind.

- Place treats in both of your hands.
- Show your dog one of the treats, close your fist around it, and present your fist to them. Give the command "Leave it." Of course, your dog won't leave it alone!
- If your dog tries to sniff, paw, lick, bark, or mouth your fist to get the treat, ignore these behaviors. Remember not to give attention to negative behaviors!

- Once your dog stops trying to get the treat, give them the treat from your other hand.
- Keep practicing this until your dog immediately moves away from your fist with a treat when you give the command.
- Once your dog masters this, you can improve it by only giving them a treat when they move away and look up at you.

Once you see that your dog has gained mastery of this command, you can modify the steps to take things up a notch.

- For this part, you need two kinds of treats: a "standard" treat and a high-value treat.
- Place the standard treat on the floor, cover it using your hand, and say the command "Leave it." Since this is a variation of the normal steps, your dog might show interest in the treat on the floor.
- If your dog tries to sniff, paw, lick, bark or mouth your hand to get the treat, ignore these behaviors. Once your dog stops trying to get the treat and looks up at you, take your hand off the treat, remove it, and give your dog the high-value treat right away.
- After this, you can take it even further. Place the standard treat on the floor while only covering it partially with your hand while you say the command "Leave it."
- Again, if your dog tries to sniff, paw, lick, bark, or mouth to get the treat, ignore these behaviors.
- Once your dog stops trying to get the treat and looks up at you, uncover the treat, remove it, and give your dog the high-value treat right away.

As time goes by, uncover the treat on the floor little by little until you have uncovered it completely, and your dog doesn't try to get it when you give the command. Upon mastery, you can try doing this while you are standing up. For this, repeat the same steps—but if your dog tries to get the treat, use your foot to cover it up while saying "No."

Get off

This command is essential to keep your dog off the furniture or off the bed. While it's not as common as the other commands, it's still considered a basic command because you may have to do it constantly, especially if you want to prevent getting pet hair on your furniture.

- If you see your pooch on your bed, sofa or any other piece of furniture, say the command "Get off" or simply "Off" while encouraging them to approach you.
- If your dog gets off, calmly reward then with a treat and some praise. If not, physically guide your dog off while saying the command.

For this command, teach the other members of your household to do it too. Consistency is key, especially for this particular command. So, when they see your dog sitting or lying in a place they aren't supposed to be, your family members should also give the "Get off" command.

Chapter 3: Training Puppies

When you take a puppy home, they won't come with an understanding of your rules or your language. A puppy is like a blank slate with natural instincts and behaviors. If you want your puppy to grow up to become a well-behaved dog, you must put in the time and effort to train them. Within the first month of bringing your puppy home, consult with a veterinarian for a wellness exam. This gives you a chance to get the first or second round of vaccinations for your pup, depending on whether your little dog has already had vaccinations or not.

Each time you go to the vet with your puppy, bring high-value treats to calm them down and make them feel happier. After the first round of vaccines, you can start off with your puppy training. Widen your pup's world a bit more by taking them to safe locations for socialization. Some examples of such places are dog training centers (make sure that these are clean and well-managed), local shops or restaurants, your neighbor's house, and public parks. However, if your puppy hasn't had their complete vaccinations yet, stay away from places that have a high likelihood of dog traffic. When it comes to puppies, you have up to 12 weeks to provide them with positive exposure to and experiences with the human world as this is when their brains are still developing.

Since you are likely to be bringing your puppy home when they are about 8 weeks old, that only gives you about 4 weeks to provide as much positive exposure as possible. The very first month of your puppy's life with you is when you should gradually introduce them to different kinds of people, different environments, other dogs (both young and old), and car rides. As much as possible, introduce your tiny dog to several new stimuli throughout the day—at least 5 would be ideal. After introducing these new stimuli, keep exposing your puppy to them while pairing the experience with a treat or some other fun activity. Just don't force your puppy into things that make them feel anxious or scared. When you go into a new environment (a safe one, of course), allow your puppy to greet it with enthusiasm and with their natural urge to explore, run, and sniff at everything. This will help build confidence and resilience in your pup.

Another thing you must teach your puppy within the first month is that your touch—and the touch of the other members of your household—is something comforting, not threatening. If you have brought home a puppy who has been abused or one who hasn't been cared for during the first weeks of their life, convincing them that touch is a good thing may take more time. But just keep your touches gentle, affectionate, comforting, and loving so that your puppy will eventually look forward to them.

From the second to the fourth month of your puppy's life, you can start house-training them. Fortunately, around this age, house-training comes quite easily as they are naturally programmed to eliminate in one place. Make sure to prepare a safe place for your puppy to do their business—a place that smells and seems familiar to them. Throughout the process of house-training, maintaining consistency is key. Also, you may want to start indoors first before teaching your puppy to do their business outdoors. Keep the rewards coming as you house-train your dog until they learn how to do it on their own. If your puppy has any accidents while you are training them, don't punish them for it. When it comes to your puppy's bodily functions, try not to do anything that will cause a negative association with them. Just stay calm, assertive, and quiet as you bring your puppy to where they should do their business.

When it comes to puppy training, there are so many things for you to learn and teach to your tiny dog. Let's go through the fundamental and practical tips for puppy training to arm you with the knowledge you need to train your new companion successfully.

How Early Should You Start?

Just like babies, puppies are delightful bundles of joy, curiosity, and energy. But a lot of times, they can also be very frustrating and exasperating. As an owner, it is up to you to deal with all of the challenges of training puppies—and the more knowledgeable and patient you are, the shorter your adjustment period will be. This adjustment period will be a lot less stressful too.

After you bring your puppy home (able to leave their mother at 8 weeks old), allow him or her to settle into your home—which is a new environment for them—first before starting with your training. If you have taken or bought your puppy from a good breeder or shelter, they may have started the socializing part of your training already. If not, then it's time for you to catch up. For puppies, keep in mind that their attention spans are very short, so don't make your training sessions too long. But by 8 weeks old and above, you can start teaching your puppy the most basic commands (the ones we went through in the previous chapter).

When you start training your puppy at 8 weeks old, only use positive reinforcement training methods along with a lot of gentle encouragement. Schedule short training sessions for your puppy, but make sure to have them every day. Integrate your training sessions throughout the day—at least 15 minutes per session will suffice. This means that you can have 3 sessions of 5 minutes long throughout the day. If you share your home with others, you can ask them to take turns with you in training your puppy. Also, train your puppy in different locations around your home to make them feel more relaxed and comfortable as they grow up.

If you want to have a well-behaved, well-trained dog in the future, make a commitment to reinforce your puppy's training daily, especially for their first year of life. Determine the training method to use for teaching the basic commands to your dog. Once you have taught these basic commands, keep practicing them every chance you get. For instance, before you give your puppy their meal, ask them to "Sit" while you prepare the food. Or before you go out with your puppy, ask them to get "Down" while waiting for you. That way, you will be able to build a routine for your puppy to help them learn what is expected of them. When it comes to practicing the basic commands, be as creative as possible. It's helpful for you to think about the rules and routines you want your puppy to follow first. That way, you will have a guide for when you are training your puppy to learn them.

Coming Up with a Training Schedule for Your Puppy

Coming up with a training schedule for your puppy is as easy as coming up with a schedule for your tasks for the day, whether at home or at work. The important thing is to create a structured schedule so that you can teach your puppy what they should do while living in your home. For the first few weeks (and months) of your puppy's life, here are the things to include in your training schedule:

When your puppy is 8 weeks old

As previously mentioned, the earlier you start training your puppy, the better. Before you bring your puppy home, you should have already prepared yourself and your home for basic puppy training. That way, all you have to do is execute your plans. Here are some things to include in your puppy training schedule:

- **Household rules**

    It's important to teach your puppy the behaviors that are allowed in your home and those that aren't. When it comes to household rules, here are some things to include in your training and training schedule:

- Train your puppy that "Yes" means that you like the behavior they are doing.
- Train your puppy that "No" means they should stop doing the behavior.
- Train your puppy to go inside their crate and remain there quietly when you close the door.
- Train your puppy to go to bed at the same time each night (around the same time as you go to bed).
- Train your puppy to go out at the same time each morning.
- Train your puppy to know where their grooming spot is—where you brush your puppy's fur, where you clip your puppy's nails, where you clean your puppy's teeth, and so on.

- **Feeding routines**

To make it easier for your puppy to learn, make sure that their water and food bowls remain in one place. Puppies learn better with routines, meaning repetitiveness, predictability, and familiarity. Therefore, when teaching things to your puppy, try to do the same thing each day using the same order of events and the same commands or words. Here is an effective way to build a feeding routine with your puppy:

Fig. 6: Feeding. From Pixabay, by Martina Behrendt, 2016, https://pixabay.com/photos/cavalier-king-1444018/ Copyright 2016 by Martina Behrendt/Pixabay.

- When you're about to prepare the meal for your puppy,

give them a cue by asking a question like, "Do you want your FOOD now?" No matter how you compose your command, make sure to exaggerate the most important words for your puppy to remember them.
- Allow your puppy to accompany you to the kitchen, where you ask them to "Sit."
- Get your puppy's food bowl from the same place, then set it in the same place each time. As you prepare your dog's food, you will be watched intently. Following the same routine for preparation teaches your dog that you are their food source.
- If your puppy isn't sitting down or if they are acting too excitable, don't set the food bowl in front of them. Otherwise, your pup will learn that excitable behavior makes food appear! Instead, allow your puppy to calm down first before giving them the food bowl.

Teaching your puppy to sit before giving their food bowl encourages patience and calmness, two essential traits to make the process of training smoother and easier. Once your puppy has started eating, don't approach them. Tell the other members of your household to do the same. Other pets shouldn't approach your eating puppy either. If your puppy walks away without finishing the food in their bowl, make a note of this; it may be an indication that your puppy is ill. If it happens frequently, consult with a vet. After about 10 minutes, pick up your dog's bowl. This helps prevent the development of food guarding or picky eating habits.

Wrap up your puppy's feeding routine by having a potty break right after each meal. If you are in the process of housebreaking your puppy, take them out on a leash. Make sure to announce the activity first before you take your dog out by saying something like, "It's time to go OUT!" Follow this kind of feeding routine each day for your dog to learn it quickly.

- Crate-training

    Crates are an invaluable tool for puppy training as they help protect your puppy from household dangers, help you with housebreaking, and give your puppy a safe place to settle in and relax. In other words, your puppy's crate is their own secure and safe den. While your puppy might, at first, feel unhappy because their movements get restricted, the more you encourage the use of the crate, the more your puppy will get used to it. Soon, you will notice that your puppy is entering the crate on their own to rest, take naps, or simply escape from the hustle and bustle of your household.

    As soon as your puppy gets used to their crate, it will be easier for you to take them along for car rides and trips to the vet. You should also make sure to keep your puppy inside the crate when you are not there to supervise them directly. This is especially true if you haven't housebroken your puppy yet. Otherwise, you might find a lot of surprises all around your home. Crate training is important because puppies shouldn't be given too much freedom. Pups who are left loose around the house can either develop bad habits or end up getting hurt.

When your puppy is 9 weeks old

As your puppy reaches 9 weeks, continue including the previous routines in your puppy's schedule to reinforce them. Apart from these, you must also teach your puppy the following:

- **More basic rules**

    To make it easier for you and your dog to live together in harmony, you should introduce basic rules little by little. Other rules to introduce to your puppy starting when they are 9 weeks old include the following:

- Train your puppy to remain calm while indoors. This reduces the likelihood of developing behavior problems. Don't allow a lot of jumping, barking, rough play, running around, and other disruptive behaviors.
- Train your puppy not to nip or mouth at your feet or hands—or anyone else's.
- Train your puppy to gently take toys and food from your hand. If your puppy tries to grab things from you, don't give in.
- Train your puppy to be gentle by being gentle with them as well. You can also do this by being firm but gentle while you are correcting your puppy's behaviors.
- Train your puppy how to interact with other people and pets. Train your puppy restraint, especially while playing with humans and other pets.
- Train your puppy that jumping up on people—even you—isn't okay.
- Train your puppy to remain as still as possible while grooming.

- **Accept handling**

As soon as you bring your puppy home, start handling them immediately. This makes it easier for your puppy to accept the things you need to do with them. Make sure that your puppy accepts the fact that you are the leader, and therefore, you make the decisions about which behaviors are okay and which behaviors aren't.

When you are handling your puppy, do so gently. The more you handle your puppy, the more they will come to accept that everything you are doing is for their own good. As part of handling, you must also train your dog to respect any other pets you have at home. Your puppy shouldn't take anything away from other pets nor should they show jealousy, pushiness, bickering, pestering, or other negative behaviors.

When your puppy is 10 weeks old

At 10 weeks old, continue practicing the training and commands you have already introduced. Apart from these, include the following in your training schedule:

- Train your puppy to learn how to walk with a leash without pulling you. If your dog continues pulling on the leash when you take walks, train them to stop this behavior first before you continue taking walks.
- Train your puppy to wait in front of open gates and doors until you give them permission to go through.
- Train your puppy to come when you call. This is one of the more important commands to teach your puppy.
- Train your puppy to bark only when appropriate.

When your puppy is 12 weeks old

When your puppy has reached 12 weeks of age, continue with your training. By this time, your puppy should already be well on their way to becoming a well-rounded dog. But you're not done yet; there is still a lot to learn. As you continue with your puppy training, you want to include the following:

- Train your puppy to remain seated until you have given a command for them to get up.
- Train your puppy to get into their crate or bed when you give the command and remain there until you give them permission to get up.

Basically, this is the time when you train your dog to control their impulses, to remain calm, and to encourage both mental and physical relaxation. If you have been training your puppy since the beginning, your pup will be able to learn these things more easily.

When your puppy is 16 weeks old

At 16 weeks old, keep on training and reinforcing the things your puppy has already learned. This is also the time when you can introduce the following:

- Train your puppy to pay attention to you and stay close to you while you take structured walks.
- Train your puppy to either greet other people and animals politely or simply ignore them. If your puppy acts aggressively, fearfully or excitably, don't allow this.
- Train your puppy to play with different kinds of toys. As much as possible, introduce a variety of toys to your puppy to determine which types they like the most and which ones they aren't too keen on. Toys keep your puppy busy, stimulated, and happy. They can also be used as part of positive reinforcement training methods.
- Potty-train your puppy. This is one of the most challenging things to teach young dogs, but it is essential. Create a separate schedule for potty-training your dog wherein you take them out as soon as they wake up in the morning, after meals, after nap time, and right before going to bed. It's much easier to potty train puppies when you create and stick to a potty-training schedule.

Now that you know more about what to include in your puppy's training schedule, you can start creating your own. But even if you make a schedule, if you see that something isn't working or you think that your puppy isn't ready to learn some of the things we have discussed here, make adjustments as needed. Don't force your puppy to do things they aren't ready for as this might make training a negative experience for them.

Potty-Training Your Puppy

Potty-training puppies requires a lot of patience, positive reinforcement, and consistency. Your main goal here is to bond with your pup while teaching them good habits. Normally, it would take between 4 and 6 months before you can fully potty-train your puppy. For some puppies, it may take longer. In particular, smaller dog breeds have higher metabolisms and smaller bladders, meaning that you will have to take them outside more frequently.

Another factor that might have an effect on the length of time you would be able to start potty-training your puppy is their past living conditions. If during their first 8 weeks of life, your puppy was never introduced to the concept of potty-training, you will first have to break their old habits before teaching new ones. If you experience any setbacks while training, don't let them bring you down. Just maintain your potty-training program to provide your pup with the consistency they need to learn. Also, keep the praises and rewards coming so that your puppy will always feel motivated to follow you.

According to experts, you may start potty-training your pup when he or she reaches the age of 12 weeks. By this age, puppies already have sufficient control of their bowel movements and bladder, making it easier for them to learn how to hold it in. If you bring home a puppy who is over 12 weeks old and who is used to eliminating inside their crate or cage, expect the process to take longer. But you can always reshape the behavior of your puppy with rewards and encouragement.

When it comes to potty-training puppies, a lot of pet owners start training indoors. This is especially true for those who don't have yards or for puppies who haven't received all of their shots yet. Start potty-training indoors then gradually transition your puppy to going outdoors.

For indoor potty-training, one of the best ways to train your puppy is with the use of puppy pads. First, determine the space where you will start potty-training your pup. A laundry room, bathroom or any other confined area with floors that are easy to clean would be ideal. Also, make sure that you puppy-proof the space first to ensure the safety of your little dog. Then, set up a space on the floor by covering it with pee pads and placing the bed of your puppy in the corner. Here are some steps to guide you as you start your puppy's potty-training routine:

- Change the pee pads regularly. But it's also helpful to place a small piece of a soiled pad atop the padded area where you want your pup to pee. The purpose of this is to provide your puppy with a scent to remind them that this is the area where they have to go potty. Change the soiled pad regularly too so that it doesn't start decomposing on the spot.
- When you notice that your puppy has learned how to go potty in one area, you can start removing the pads right next to their bed.
- Over time, you can gradually remove more and more pee pads until you are left with 1 or 2 sheets placed in the area where your puppy usually goes potty on.
- If your puppy constantly goes potty on the remaining pee pads, you may extend their areas of access gradually. But if your puppy starts having accidents, reduce that area once again.

After successfully potty-training your puppy indoors, it's time to start training them to do their business outside. To do this, the first thing you need to teach your puppy is the "potty cue." This is an important part of the training as it serves as a signal to go potty. Your goal here is to show your pet how to use the potty cue to tell you that they need to go outside. Here are some steps to help you transition from going potty indoors to going potty outdoors:

- **Teach the potty cue to your puppy**

First, bring your puppy by the door and give them the command to "Sit." Then, you can either teach your puppy to bark, ring a bell, or scratch the door if they want to go out. At first, you may have to demonstrate this. The potty cue is simply a signal that your puppy wants to go out to do their business, so you shouldn't use it when you're just taking your pup out to play. When your puppy performs the cue you taught, open the door, let them out, and give them a treat. You can think of your own potty cue. When it comes to this signal, you decide what you want it to be.

- **Determine your puppy's potty area**

    Before you take your dog out, place them on a leash. This is an especially important step so that your dog knows that after the potty cue, taking them outside has a specific purpose. Take your puppy to the area that you have set as their potty area and stop there. Patiently wait until your puppy has done their business. After that, reward your puppy with treats and praises. This makes the whole experience a positive one so that your puppy will look forward to doing it again. But if your puppy doesn't do their business, bring your puppy back inside your house, then repeat the process. Over time, your puppy will learn what these actions mean.

- **When your puppy is left alone at home, leave them in their crate**

    Unless you have fully potty-trained your puppy, you should keep them inside their crate when leaving them home alone. This will help prevent accidents in the different parts of your home while you aren't there. If they were to have an accident while they were home alone, you wouldn't be able to correct your puppy's behavior.

As you build and strengthen your puppy's potty routine, make sure to show your pup a lot of love and encouragement. Don't make the experience a negative or scary one as this might end up teaching your puppy the wrong things. Also, don't punish your puppy for not learning things as quickly as you want them to. Simply correct the behaviors as needed. Remember... patience is key!

Practical Tips for Training Puppies

Fig. 7: Puppy Training. From Unsplash, by Jairo Alzate, 2015, https://unsplash.com/photos/sssxyuZape8/ Copyright 2015 by Jairo Alzate/Unsplash.

Starting off on the right foot (or paw) with your puppy makes them feel more secure in your home and in your presence. To continue providing good experiences to your pup, use positive reinforcement methods as a foundation for your puppy training. Consistent, gentle, and loving puppy training provides you with a strong foundation for your relationship with your puppy. When you start early, you will be able to live with your puppy peacefully and without (much) stress. Teaching your dog to learn what is expected of them from the beginning will have a positive impact on the healthy growth and development of your young canine companion. To wrap up this chapter on puppy training, here are some more practical tips for you to use:

- **Research how puppies learn**

    Reading this book about dog training is an excellent way to learn more about dogs and learning. When you have a puppy, you can think of them as a clean slate. With the proper guidance and learning techniques, your puppy can become whatever you want them to be. To reinforce what you learn here, you can also talk to professional dog trainers and ask them for advice. Learning everything you can about training puppies and dogs of all ages equips you with the knowledge you need to ensure training success.

- **Set clear boundaries from the start**

    No matter how cute your puppy is, you should set clear boundaries for them and enforce these boundaries. Of course, this doesn't mean that you should be overly strict with your puppy. Train, educate, and be as firm as needed, but whenever your puppy does a good job or learns the behavior you want them to learn, shower your pup with love, praises, and treats.

- **Make sure you give signals explicitly and commands clearly**

    Puppies are very good when it comes to reading your facial expressions. They will often react quickly when you give simple commands or simple hand signals compared to complex sentences or signals. This is why the basic commands only consist of 1 or 2 words. It's much easier for your puppy to learn these commands when you make them as simple, as clear, and as explicit as possible.

- **Develop your puppy's social skills early on**

    For the first five to six months of your puppy's life, try to introduce them to a hundred different people or so. This makes your puppy comfortable in the company of others, not just you. To make this even better, bring a small bag filled with your puppy's favorite treats with you and ask the people your puppy meets to feed them treats. This teaches your puppy not to feel threatened by other human beings.

If you have the resources, you may also consider enrolling your puppy in kindergarten class so that they can play with other dogs while learning essential social skills. Puppies should learn how to interact with all other living things around them for them to grow up to be happy, healthy, and well-balanced.

- **When you take your puppy to new places, make it a fun experience**

    While your puppy is still young, don't focus too much on obedience training; you can do this type of training later on. But when it comes to socialization and exploration, these must be done while your puppy is still young. They will have an impact on how your puppy's temperament will be when they grow up. When taking your puppy to new places to meet other people or animals, make each experience fun. This will help prevent the development of fear or aggression issues along with other kinds of behavioral problems when your puppy grows into an adult.

- **Your puppy training should include walking on a leash**

    Obvious as this tip might seem, many dog owners neglect this until their puppies have already grown into young or adult dogs. But training your puppy to respectfully and calmly walk on a leash can help you with other training methods, such as potty-training and socializing.

- **Understand your puppy's natural guarding instinct**

    Puppies are naturally inclined to protect their toys, food, friends, and everything else they hold dear. When you notice your puppy having fun with something or someone, taking this away will teach your puppy to guard them instead of giving them up. If you want your puppy to let go of something, distract them with another object before you take it away.

- **Don't use your puppy's crate as punishment**

    You want your puppy to love their crate and feel comfortable in it. But once you start using it as a form of punishment, your puppy will hate the crate—and this is very difficult to unlearn. Using any form of punishment (this differs from discipline, which we will be going through later on) might have detrimental effects on your training, especially when it comes to puppies.

- **Your puppy training should include learning not to bite, nip, or mouth**

    While your puppy is still young, teach them that biting isn't okay. If your puppy nips at you or bites you a bit too hard, say "Ouch!" using the same pitch as a puppy's yelp when their tail or foot is stepped on. This serves as a warning that what your puppy did wasn't a good thing. If you see that your puppy listens (meaning they stop biting you), give them a treat or praise them. You can also choose to ignore the behavior, then turn around and tuck your hands under your armpits. This is a calming signal for puppies and dogs.

- **Gain your puppy's trust so that you can establish a strong bond**

    As much as you love your puppy, he or she should also love and trust you in return. But you cannot simply demand this love and trust; you must work hard to earn it. Once you have gained their trust, it will serve as the foundation of your bond. The more you treat your puppy the right way, the stronger your bond becomes. With this strong bond, your puppy will want to please you all the time, thus making training easier.

- **Encourage your puppy to explore**

  The more you allow and encourage your puppy to explore their surroundings, the more confident they become. You can also encourage your puppy to explore different objects both at home and in different surroundings. Finally, encourage your puppy to explore and interact with people and other dogs (even other pets if you have them at home). All of this exploration helps your puppy develop into a well-rounded and well-balanced dog.

- **Your puppy training should include teaching your puppy how to stay home by themselves**

  When you leave your puppy at home, they will feel very stressed. Of course, you can't bring your pup everywhere you go, right? Therefore, you should train your puppy to remain inside their crate while you are out. This is why it's important to get your puppy used to their crate. The more relaxed and comfortable they are, the easier it will be to leave your puppy home alone. You may also want to leave a toy in your puppy's crate to keep them entertained. Just make sure that the toy you leave is something your puppy is interested in and is something safe.

  Positive reinforcement works when you're trying to train your puppy to stay home alone. You can place your puppy's crate in a room, then leave them there for some time. If your puppy doesn't make a fuss while you left them alone, reward them for it. Then, you can increase the length of time that you leave your puppy to help them learn that they will be okay even if you're not around.

- **Introduce new words to your puppy and keep expanding their vocabulary**

  As previously mentioned, puppies aren't born with an

understanding of English vocabulary (or any other language for that matter). If you want your pup to learn what specific commands mean, it's your responsibility to teach them—and you should do this right away. Talk to your puppy frequently to strengthen your bond. Share your experiences through words. Over time, this will help your puppy learn what words mean. This might even make it easier for your puppy to learn commands once you start puppy training.

- **Provide guidance and encouragement for your puppy instead of trying to control them**

The key to successful puppy training is not to control, force, or coerce your puppy into doing what you want. Instead, provide your puppy with a lot of love, encouragement, praise, rewards, and guidance to make the experience more positive. This is especially important for puppies so that they won't grow up fearing or resenting you.

Chapter 4: Training Young Dogs

While most people prefer to adopt puppies, you may end up with a dog who is a bit older. After reading the last chapter, you might be thinking, "How do I go about this now?" The good news is that when it comes to training dogs, you can start at any age and stage!

Before you bring your young dog home, try to find out all you can about their earlier weeks or months of life. Did your young dog experience any prior training? Was your young dog able to socialize and explore adequately? Has potty or crate-training been introduced to your young dog? These are important questions to ask your source to determine where you should start with your training.

Ideally, though, whether your young dog has already experienced training in the past, you will want to start at the beginning. Your young dog will be used to the routines of the place where they grew up. But you will have your own routines at home, and your dog must learn these. To catch up with your dog's training, come up with a training schedule for each day and stick with it, no excuses.

You may want to start with potty-training as this would already be considered "delayed." Follow the same steps that we have discussed in the previous chapter for potty-training puppies. Depending on the past experiences your young dog had in terms of training, this may either be very easy or very challenging. But this is one skill you must train your dog to do as part of their discipline. Then, you can start expanding your dog's vocabulary, teaching the basic commands, and more. Also, don't forget to shower your dog with love, affection, encouragement, and rewards to make your young dog happy, healthy, and well-adjusted.

Learning from Older Dogs

Have you ever heard of "allelomimetic behaviors?"

It sounds strange, doesn't it?

But this concept is actually very important when it comes to other dogs. Allelomimetic behaviors are a type of group-coordinated behavior. They depend on dogs' natural inclination to follow, imitate, and want to be in the company of other dogs. You can use this knowledge to help your young dog (or even your puppies) learn the proper behaviors.

Fig. 8: Dogs Learning. From Pixabay, by Spiritze, 2016, https://pixabay.com/photos/dogs-puppy-cute-animal-pet-canine-1790046/ Copyright 2016 by Spiritze/Pixabay.

Allowing your young dog to socialize with older, well-trained dogs will help them learn a lot of essential behaviors, especially those that are socially significant.

If this is your first time attempting to train a dog, and you have a young dog, you can reinforce, simplify, or improve your training by allowing them to spend time with older dogs. Once your young dog has gotten used to these older dogs, you will start noticing the allelomimetic behaviors. For instance, if the owner of the older dog tells their dog to "Come," the dog will obey... and trailing behind that dog will be your own. In the same way, when your young dog is able to watch older dogs being trained, they will have a better idea of what is expected of them once you start with their training.

This type of learning can be extremely effective when done right. If you want your young dog to learn from older dogs, find friends or family members who have already trained their own dogs successfully. Then you can start scheduling "play dates" for your dogs for the purpose of enjoyment, socialization, and education. The more you find ways to make dog training easier, the higher your chance of success will be. As far as ease and convenience go, this method is a winner!

Top Tips for Training Young Dogs

As challenging as dog training is, it can also be one of the most rewarding and fun aspects of raising a young dog. As soon as you bring your small dog home, make dog training one of your top priorities. Dogs are very trainable, even if they've grown beyond the puppy stage. Even professional dog trainers will accept dogs of all ages, breeds, and abilities. Of course, we have already discussed the added benefits of training your dog on your own. It helps strengthen your bond with your dog, and it also helps you learn more about your dog too.

For dog training, the most effective methods are those based on rewards or positive reinforcement. When you give your young dog toys, praise, or treats that they really like, this will help make your training much easier. Since you will be training your dog yourself, you will be able to determine exactly what motivates them. Yes, you will experience struggles and failures while training your dog, but don't let that stop you. Keep moving forward by learning more about your dog and using these helpful tips for training your young one:

1. **When setting and enforcing rules, maintain consistency**

    It cannot be said enough: consistency is extremely important when it comes to dog training, especially in terms of setting rules and enforcing them. Make sure that each day, every day, your dog follows the rules that you have set. If there are other people living in your home, make sure that they enforce these rules too.

This means that you may have to teach them how to train your young dog (or let them read this book too!) so that they know what needs to be done. While you will be doing the introductions and most of the teaching, the other members of your household should help with the reinforcement. This consistency will help your young dog understand their role at home and what is considered acceptable and unacceptable behavior.

2. **Maintain consistency with your words too**

While young dogs may already understand a few words, this doesn't mean that they have an adequate vocabulary already. Therefore, it's your responsibility to teach them. You may talk to your dog about things around you and the experiences you have. But when it comes to training and teaching commands, use the same words over and over again. For instance, when you're training your dog to "Sit," don't say "Sit" one time, then "Sit down" the next. This will just end up confusing your canine companion.

3. **Be aware of the limitations of your young dog**

The best way to make yourself aware of your dog's limitations (and abilities) is by learning about their past from your source. If you acquired your dog from a shelter or a breeder, you can interview the people who took care of your young dog first. Then, compare what you have learned with what you observe while interacting with and training your dog.

On the other hand, if you acquired your dog from a rescue shelter or you decided to adopt a stray, you may have to do a lot of observing and experimenting to determine your dog's limitations. Either way, learning what your dog can and cannot do will help you customize your training methods.

4. Encourage and reinforce the good behaviors of your young dog

If you want to have a well-behaved, happy dog, shower them with encouragement, praise, rewards, and other good things. This is why positive reinforcement training methods are recommended instead of methods that incorporate harsh words and punishments. Think about it; when you constantly shout at your dog and hit them, how will they learn good behaviors?

While you are observing your dog, praise them for doing good things no matter how small or ordinary those good things are. Doing this will help reinforce their behavior when you are training your dog. When you praise, encourage, or reward your dog during training, they will remember the times when you did the same thing. Over time, your young dog will start learning the behaviors that bring about these reactions from you.

5. Consider your young dog's environment while training

No matter what age your dog is, you must always consider the environment where you are training them. This is especially important when you are introducing new commands or tricks. Perform your training in an environment free of distractions to improve your success. Conversely, if you try to train in a new environment or in a place that makes your dog excited, you might see your training session go to waste. Think about the training you have planned and choose the appropriate environment for it.

6. Don't expect to be a perfect dog trainer

For a lot of dog trainers, when they make mistakes, they stop dead in their tracks. They take this as a sign that they aren't meant to train their dog or that they don't have the right skills for the task. Don't be too hard on yourself. If you make mistakes, learn from them. Make adjustments to your training methods to prevent those mistakes from happening again. There is no such thing as a perfect dog trainer, not even among professionals. We all make mistakes, we all have our triumphs, and we can always learn new things along the way. If you feel like you're struggling, try to learn more about your young dog. You can also ask for help. Just don't give up on dog training entirely!

7. **Customize your training methods to suit the needs and abilities of your young dog**

The great thing about dog training is that it's completely flexible. Although there are basic commands and effective ways to teach them, you can customize your training method to suit your own individual dog. For instance, while treats work for most dogs, your young dog might prefer praise or toys. So, even if you give them treats, they won't be motivated. Try to determine what your dog likes, what works, what doesn't work, and adjust your plans accordingly. This may seem intimidating at first, but over time, you will become a master of knowing how to train your dog in the best possible way.

8. **More training tips for you:**

Apart from these tips, you can also apply the puppy training tips you have learned in the previous chapter. Depending on what your young dog already knows and what else you want to teach them, you may have to combine different methods to reach the goals you want your dog to achieve. Here are a few more tips for you:

- o Whether your dog already came with a name or you gave them one, use it consistently. That way, your dog will learn what their name is, and when you call it out, they

will respond to you.
- Patience is a virtue; always remember that. If you feel like you're not getting anywhere with your training session, stop. Do the same if you start feeling annoyed or frustrated with your dog. Also, keep in mind that dogs learn differently and at different rates. Remembering this fact may help extend your patience.
- Apart from the words you use, your dog can also understand your body movements and voice tone. Keep this in mind while training.
- Set short training sessions each day. This will prevent your dog from getting bored while training.
- Groom, stroke, pet, and handle your dog each day so that they get used to and accept being handled.
- Learning about and understanding your dog will help you predict their future behaviors and anticipate what they will do next.
- End all of your training sessions with a command or trick your dog already knows. This ensures that your training sessions always end on a positive note.

Chapter 5: Training Adult Dogs

Have you ever considered adopting an adult dog? A lot of people prefer adopting puppies and young dogs as they feel intimidated at the thought of adopting a dog that is all grown up—and the thing that intimidates them the most is the fact that they would have to train that grown-up dog. If this situation sounds familiar to you, don't let your fear stop you! There are so many adult dogs out there without homes, and the older they become, the closer they get to being euthanized... and that is just sad.

Even if you bring an adult dog into your home, it is entirely possible for you to train them. You might simply have to take a different approach than you would with puppies and young canines. But when it comes to the principles, training adult dogs is pretty much the same as training younger ones. And the older your dog is, the more patience you will need to train them. This is especially true if you have adopted a dog who has been abused, one who has never been trained, or one who has learned wrong or inappropriate behaviors. In some cases, you will first have to train your furry friend to unlearn the behaviors they have learned before you can start training them to do what's right. To start off, perform the following training with your dog:

Crate-Training

This type of training is essential, especially if you want house-training to become an easier task for you. Crate-training is a helpful and effective way to begin house-training your dog as it works with the natural instinct of canines not to do their business in their den. For most adult dogs, when they consider their crate as their den, they will try their best to "hold it in" as long as they are inside. But as soon as you notice that your dog needs to go, let them out immediately. We will go through more detailed steps for house-training your dog in the next section, but for now, here are some helpful pointers to keep in mind as you train your adult dog to get comfortable with their crate:

- Select a crate that your dog will fit into comfortably and one that you can clean and maintain easily.
- Train your dog to learn that they cannot access all the rooms of your home.
- Help your dog become more comfortable in their crate by establishing a routine that you follow each day.
- Take your dog outside for potty breaks. Do this about 4 to 5 times each day. If your dog doesn't do their business when you take them outdoors, that's okay. Just observe your dog carefully when you go back inside.
- Stock up on products for cleaning and removing stains of dog urine so that you can maintain the cleanliness of your dog's crate.
- If you aren't at home or you need to do something important while you are home, leave your dog in their crate.
- Observe your dog's behavior while inside to crate. This will help you determine whether your dog is merely whining to complain about being "trapped" inside the crate or they are truly feeling anxious about being left inside the crate on their own.
- To ease your dog's stress or anxiety, place 1 or 2 toys (preferably the indestructible ones) in with them to keep them entertained.

Be patient with crate-training and never force your dog to go inside their crate. Over time, you may start noticing that your dog goes into the crate even without prompting or rewards. When this happens, it means that they already feel comfortable in it.

Socialization

Important as this aspect of dog training is, a lot of dog owners end up neglecting it. This part of the training involves helping your dog feel comfortable in different kinds of social situations. This may be one of the more challenging things you will have to do if you adopt a rescue dog or one with a painful past. But if you are able to socialize your dog properly, you can transform your dog into a confident, friendly, and happy canine. Just make sure that you continue with socialization throughout your dog's lifetime to make it effective.

When it comes to socializing your dog, take things slow. As much as possible, follow your dog's pace and don't force them to interact or socialize with others right away. Also, don't attempt to push your canine out of their comfort zone at the beginning. You don't want to end up making your dog more fearful or anxious, right? As with other types of training, keep in mind that not all dogs learn in the same way or at the same rate. Your dog may find some people or situations easy to adjust to while they might also dislike some people or will refuse to go into certain environments.

Consider your dog's personality and past life too. If you obtained your canine from a breeder who cared for your dog well and trained them adequately, you may notice that socialization comes easily to them. On the other hand, if you obtained a dog from a shelter that saves dogs who have been abused or abandoned, you may notice that socialization isn't an easy thing to get used to.

Leash Training

Fig. 9: Leash Training. From Pixabay, by Katrin B, 2015, https://pixabay.com/photos/dog-school-dog-training-rottweiler-672716/ Copyright 2015 by Katrin B/Pixabay.

There is one main difference that exists between leash-training adult dogs and leash-training puppies, and that is the amount of pain you will experience in your arm! Naturally, when a puppy pulls on a leash, the strain you will experience won't be as much as when an adult dog pulls on a leash, especially if you have chosen one of the bigger dog breeds. But when it comes to the principles and methods of leash training, these remain the same across all ages. Here are some tips to help you out:

- Prepare to correct your dog at each step of the process and try to increase your patience. It will be difficult, but not impossible.
- It's best to begin leash-training your dog at home before you take them to the park or to any other public place.
- If you want to gain control of your dog more effectively, purchase a high-quality training harness or collar for your canine companion. Just stay away from retractable leashes, as these have a tendency to be unpredictable.
- If you have a small or medium-sized canine, it would be best to use a simple collar with a chain-link design. But if you have a large or extra-large canine, it would be best to use a prong collar.
- If you have chosen to use a training collar, the right way to correct your dog is to give a "pop"—a sharp, short tug on your dog's leash to tighten their collar for a moment before releasing it just as quickly—on your dog's collar while simultaneously giving a correction verbally.

- If you have chosen to use a prong collar, make the "pop" less sharp and shorter as this type of collar requires less pressure.
- If you have chosen to use a halter or harness, all you need to do is grip the leash firmly and apply gentle pressure for correction.
- When you are leash-training your dog, never yank at their neck or drag them forcefully because you might end up hurting them.

These are just some general tips for leash-training your adult dog. But when it comes to the actual training, here are some steps you can follow:

- Hold your dog's leash with one hand while your dog is on the opposite side of your body. In this starting position, your dog's leash should be running across your front while you also hold part of the leash with your other hand approximately at your thigh level.
- Give the command to "Sit" before you begin your walk. If you haven't taught this command yet, then simply allow your dog to stand next to you.
- Stay in this position for a few seconds to train your dog that it's time to take a walk. If your dog tries to lunge or walk forward, just give them a firm but gentle pop while saying "No" or "Wait." Only when your dog is able to stay in this position for at least 10 seconds (the longer, the better!), then you can take a step forward while giving the command, "Let's go."
- Take a short walk around your yard to give your dog practice. Whenever your dog lunges, yanks, pulls, or twists, stop walking while using verbal corrections simultaneously with pops to bring your dog back to you. Only when your dog stands still should you continue walking.

Perform leash-training with your dog daily. This will help your dog learn the correct behaviors for walking with you. When you see that your dog has learned how to walk on a leash properly, you may start experimenting by taking them on walks in other environments.

House-Training Adult Dogs

Depending on where you obtained your adult dog and how they were raised/trained until the time you brought the dog home, house-training may either be an easy or difficult task. Of course, a dog who has been trained in the past will only have to be taught where to do their business in or outside of your home. But if the dog you adopted has no prior training, then you will have to start from the very beginning as you would if you chose to adopt a puppy. For this aspect of training, here are the basic things to keep in mind:

Fig. 10: Housetraining. From Pixabay, by Pezibear, 2018, https://pixabay.com/photos/dog-small-small-dog-business-3367901/Copyright 2018 by Pezibear/Pixabay.

- **Observe your dog's behaviors first**

    From the time you bring your dog home, observe their potty-related behaviors. Try to see how frequently your dog does their business, the location where they usually do their business (if your dog only does their business in specific parts of the house), or if your dog only does their business under very specific conditions. For instance, your dog might pee when you call out their name in a firm way. This may indicate that they do this out of fear. Observing your dog will help you determine how to approach house-training.

- **Rule out medical conditions**

    If you notice that your dog does their business at inappropriate times or frequently does their business indoors as if they don't have control, consult with a vet first. Do this to rule out any medical conditions that might be causing your dog's potty-related behaviors. If your dog suffers from any medical condition, you can work with your vet to help them overcome it. If not, then you can start house-training your dog.

Make sure to include house-training in your dog training schedule. Do this as soon as you bring your dog home. Start observing, rule out medical conditions, then start your house-training. Here are some practical tips to do this successfully:

- Unless you actually caught your dog doing their business or having accidents anywhere in your house, don't punish them for it.
- If you catch your dog in the act, startle them with a clap or a shout, then bring them outdoors. Once your dog finishes their business, give them a reward or praise.
- Also, make sure to clean the mess your dog made in the house thoroughly so that they don't get attracted to the residual smell of urine or poop.
- Set a single area for your dog to do their business. Each time you bring your dog outside to pee or poop, bring them to that same place.
- When you bring your dog out, don't distract them with a lot of talking or by playing games with them. Establish a routine wherein your dog will learn that you are taking them outside for a single purpose—to do their business.
- Praise your dog when they do their business outside properly. You can also bring treats with you when you take your dog outside to reward them as soon as they are finished. This helps reinforce the behavior of pooping or peeing outdoors.
- Take your dog outdoors regularly to do their business. This is especially true when your dog wakes up in the morning, after

each meal, and before going to bed each night.

When it comes to house-training, consistency is extremely important. If you aren't home, make sure that the other members of your household know that they need to take your dog out at regular times throughout the day. When you see that your dog is already getting used to doing their business outside, you can teach them the potty cue.

Basic Obedience Training for Adult Dogs

Basic obedience training involves teaching your dogs the basic commands. These basic commands are called "basic" because they are the easiest to teach dogs, and you can use them when you want to introduce more advanced commands or tricks to your canine companion. Also, basic obedience training makes it easier for you to keep yourself and your dog safe in different kinds of situations.

While you are house-training your dog, you can also start their obedience training. For this, you need a room or space in your home that is free of distractions. Start in such an environment to introduce the commands and practice them a couple of times before moving on to other environments that may contain distractions. Also, make sure that you have enough rewards (like treats or toys) to give your dog one each time they do the command correctly.

When it comes to obedience training, don't expect your dog to learn things in one go. Again, this will depend on a number of factors, such as if your dog has been taught basic commands in the past, the temperament of your dog, the personality of your dog, and even your dog's breed! The simplicity of the command is another factor that will affect the rate at which your dog learns. Here are other pointers for you:

- Maintain consistency throughout your obedience training. This means that you follow the same routine, stick with the same rules, and use the same commands each and every time.

- Dog training is something you will have to do for the rest of your dog's life. And each time you have a training session, make sure to bring a lot of patience with you. This is especially true at the beginning when you are just introducing new commands or tricks.
- Don't give your dog time-outs because they don't understand what these mean. Just correct your dog's behavior right away as soon as you catch them in the act and praise them immediately when they do things correctly.

Basically, to increase the success of your dog's obedience training, make the experience more positive. Set your dog up for success to keep them happy and enthused. This will make your dog training a lot easier and more enjoyable for both yourself and your furry friend.

Adult Dog Training Tips from Experts

Small or big, young or old, dog training is an important part of your dog's life. When you have made the decision to make an adult dog part of your life, keep in mind that you're handling the task of training a dog who has already had a lot of experiences in their life. Therefore, you will be contributing to the lifelong training process that you will continue until your dog grows old. Before you start training, take the time to learn more about your dog in order to understand them better. This will help you determine the best approach to your dog training.

Also, be as realistic as possible when it comes to your dog's training. While it is possible to train adult dogs and teach them new tricks, the process might take more time, effort, and patience than what you expect. No matter how slow your dog's progress is, don't let it discourage you. Invest in your dog, and in time, you will see the fruits of your labor. It's also important to establish a strong bond with your canine. Spend a lot of time with your dog, shower them with affection, and reward them whenever they do something good. These actions will go a long way in motivating your dog. Here are more tips for you from various dog training experts:

- When your dog has already gained mastery of the basic commands, you can move your training sessions to different environments. This will help your canine understand that they should follow you and the commands you give no matter where you are.
- Apart from your training sessions, reinforce the commands you teach by using them throughout the day. For instance, you can ask your dog to "Stay" in preparation for a walk. Or you can ask your dog to "Sit" in the kitchen as you prepare their meal. Using the commands in different situations reinforces them and strengthens the routines you are trying to build.
- Focus on positive reinforcement training methods wherein you use a lot of praise, rewards, and affection while teaching your dog new things. Avoid talking to your dog harshly or using physical abuse as these might have adverse effects on your canine.
- Learn the importance of the basic commands and why you are teaching them. For instance, commands like "Sit" or "Stay" help control the sudden impulses or urges of your dog.
- Instead of expecting your dog to sit, lie down, or calm down all day, help your dog manage their energy. To do this, you must provide your dog with enough mental and physical stimulation to keep them occupied throughout the day. Training is a great way to do this, along with giving your dog toys to play with.

Instead of thinking of dog training as a chore, think of it as spending quality time with your canine companion. Each time you have a training session, this helps enhance your relationship with your dog. As your dog's behavior improves, you will also come to realize that you know them a lot better too. Have fun with training, and it will become a lot easier, both for you and your pet.

Training Adult Dogs to Do New Tricks

It is true that dogs can learn new things no matter what age they are. Some say it's easier to train adult dogs than puppies, while others say the opposite. But you won't really prove this either way unless you have tried training dogs at different ages. Of course, the important thing is the training itself. If you have just brought home an adult dog, training will maintain the sharpness of their mind and provide them with the structure they need to live in the same home as you.

One good thing about adult dogs is that they may have longer attention spans than puppies. They can also adapt quickly to changes in their environment and to new routines compared to puppies. This means that as long as you maintain consistency, you can expect to see the results you desire. For instance, in terms of potty or house-training, adult dogs have more control over their bladder and bowel movements than puppies. And if you have obtained a dog who already has previous house-training experience, then this part of your dog's training will surely be a breeze.

In some cases, you might notice that treats aren't really motivating your adult dog. In such a case, you can focus more on praises and on demonstrating the behaviors for your dog. Then, practice the behavior several times so that your canine starts understanding what you expect of them. Basically, when it comes to adult dogs, you will have to do a lot of observation, reflection, and adjusting to your training plans. Make changes as needed, keep learning more about your dog, and soon, you will discover that your dog is starting to pick up the things you are trying to teach or train them to do!

Chapter 6: Training Senior Dogs

Puppies, young dogs, adult dogs, and now, even senior dogs?

Yes! You can train senior dogs too, whether you have chosen to adopt one or your dog has already grown old and you have now made the choice to train them. As previously mentioned, any dog of any age can be trained. It's never too late. Senior dogs without any formal or structured training still have the ability to understand the concept of obeying commands to get rewards. Even if your senior dog has learned some bad or inappropriate behaviors in the past, you can teach them to think about the behaviors that will earn them a reward and how they should respond when you give specific commands. As with training younger dogs, spending quality time with your senior dog while training them helps strengthen your bond.

As soon as you have brought your older dog home, it's recommended to begin training right away. For this task, the main things you need are patience, time, and treats. Also, before you begin, make sure that you are ready for the task of training a senior dog. For older dogs, reward-based training is one of the most effective methods to use. And if you want to help your senior dog achieve your training goals, use high-value treats to motivate them. Commit to the task, understand the fact that you will be training an older dog, and don't force your canine to do what you want.

If your senior dog grew up with you (but without proper training), you will already know their training history. But if you have decided to adopt a senior dog, then you should speak to your source before you take your dog home. That way, you can find out whether your senior dog has already experienced any kind of training in the past or not. Learning about the training history of your elderly dog is important so that you can come up with a training plan and schedule that your dog can cope with. If your source doesn't know the training history of your dog—such as if you adopt a rescue dog—then you can try to find out for yourself. Try to give your dog the basic commands to see how they will respond. If your dog follows these commands, make a note of the commands they already know. But if your dog just stares at you blankly no matter what command you give, then you may have to start from the very beginning, as you would with puppies or younger dogs.

Since you will be training a senior dog, it's important to consider their health and age while designing your training plans and schedules. Keep in mind that your dog is already old, so they might get tired easily. To make sure that you don't overexert your canine, pay attention to the common signs of exhaustion, such as frequent yawning, excessive licking, drooping ears, ignoring you, and sniffing at the ground.

Should You Train Senior Dogs?

When people adopt senior dogs, or if their dogs have already reached ripe old ages, training them might not be a priority. But even if your dog is already considered a senior, there are many benefits to training them. One of the most important benefits of training senior dogs is that it will give you lots of opportunities to spend time with your aged, fragile canine. If you think that your dog only has a few more years to spend with you, then you should spend as much time as possible with them, right?

Fig. 11: Senior Dog. From Pixabay, by Leo_65, 2017, https://pixabay.com/photos/dog-senior-old-grey-snout-hybrid-2514042/Copyright 2017 by Leo_65/Pixabay.

Also, training your senior dog will help you learn more about them. At several times throughout the day, you will make your dog your main priority as you follow the training session schedules you have set. Throughout the process, you will be able to give your dog a lot of praise, rewards, and love while learning things you never even knew about them. No matter what age dogs are, training helps strengthen the bond between them and their owners. And the stronger this bond is, the more you will love your dog and appreciate them.

The old adage that says that "you can't teach old dogs new tricks" isn't accurate at all. In fact, it's the exact opposite. As long as you keep your training sessions short and use fun, positive ways to motivate your dog, the experience will surely be one your dog will always look forward to. But if you're worried about the health (and age) of your dog, you may consult with a vet first to find out the best training approach to take. It's all about learning what is best for your dog first, then doing everything you can to educate them and help them become better, more well-behaved canine companions.

Things to Remember When Training Senior Dogs

Just because you have an older dog, that doesn't mean that you don't have to train them. Senior dogs are perfectly capable of being trained and of learning new things. In fact, professional dog trainers have seen dogs older than 10 years who perform admirably in training classes. So, if you have a senior dog at home, don't make their age an excuse not to provide them with fun and educational training.

When it comes to "seniority," the age at a which dog qualifies for that classification varies. In general, when small dog breeds reach the age of 12, vets consider them seniors. For bigger dog breeds, vets consider them to be seniors between the ages of 6 and 8. This means that bigger dogs become seniors earlier than smaller dogs, and they have a higher likelihood of showing the most common signs of aging compared to their smaller counterparts. Senior dogs might also show signs of mental aging; however, these are not as noticeable.

Despite all of these aging signs, dogs will always be ready and willing to take part in dog training as long as you find fun, interesting, and motivational methods for the task. Even if your vet advises you to avoid teaching tricks that involve a lot of physical activity, you can still teach simpler commands and tricks to your canine to keep them physically and mentally stimulated. When it comes to training senior dogs, here are some things to keep in mind before you start:

- **It's never too late for treats!**

    Just because you have always believed that old dogs can't learn new tricks, that doesn't mean that you shouldn't try to train your elderly dog. The key to making this easier is to stock up on different kinds of treats! If your dog grew up with you, you should already know what treats they like. If not, then you can experiment with different types of treats to see which one works best for training.

- **Commit to your dog's training by improving your own skills**

    Your dog will only be able to learn as much as you can teach them. No matter how sharp your dog is in their old age, they won't be able to train themselves. Therefore, if you want to make the most out of your senior dog's training, you should learn everything you can about it and try to improve your own skills. The more you practice with your dog, the better you—and your canine—will become.

- **Determine your dog's physical condition and whether it will affect your training**

If your senior dog is still strong and spritely, then you have nothing to worry about. However, if you notice that your canine already suffers from physical limitations, adjust your training plan accordingly. This is why it's important to consult with your vet first—so you can determine your dog's limitations or if they suffer from any medical condition that you should work around while training.

- **Determine your dog's mental or cognitive abilities and how they might affect your training**

    Sadly, some dogs may develop "Doggie Alzheimer's" when they reach old age. Such a condition will significantly limit your dog's ability to learn new behaviors or tricks. If your dog suffers from this condition, you may notice a number of symptoms, including disorientation, restlessness, decreased hearing, barking aimlessly, losing their ability to recognize their surroundings and the people around them, and losing their previous house-training behaviors.

- **Think about the methods you plan to use for your dog's training**

    We have already gone through the different methods of dog training, and it's up to you as an owner to determine which method is best for your canine. The important thing is to focus on positive methods for training as these will encourage your canine to keep learning with enthusiasm.

- **Try focusing on one task at a time**

    You don't have to feel pressured to "catch up" in terms of dog training. This is especially true if your dog is fairly well-behaved to begin with. While training your dog, try to focus on one task at a time. This helps reduce the confusion, and training won't end up making you—or your dog—frustrated.

- **Think about what you want to teach your dog**

    If you have brought home a dog with a lot of undesirable behaviors, work on helping them unlearn these first. Do this before you try teaching new behaviors. Also, think about your training goals, what you want your dog to achieve after some time. This will help you determine what you want to teach your dog.

- **Make socialization part of your dog's training**

    If you want your dog to feel more open to training, you may consider including socialization as part of your plans. Senior dogs who haven't been trained in the past might not have been exposed to new environments, people, experiences, and animals. Once in a while, take your dog out and expose them to new stimuli to invigorate their mind and motivate them to learn more.

Of course, it goes without saying that you should keep your training sessions short but meaningful. Dog training provides your senior dog with enough challenges to keep their mind sharp and their bodies healthy. Therefore, once you feel like you're ready to take on the task, then it's time to start!

Helpful Tips for Training Senior Dogs

Fig. 12: Training Senior Dogs. From Pixabay, by fernandozhiminaicela, 2018, https://pixabay.com/photos/older-dog-old-man-with-his-dog-3568173/Copyright 2018 by fernandozhiminaicela/Pixabay.

Training senior dogs can feel more challenging and time-consuming compared to younger dogs. This is especially true for senior dogs who have had no dog training experiences in the past. But as long as you stick with your training regiment, you will be able to teach your dog the basics to make living with them a lot easier.

When training senior dogs, remember that their brains aren't "primed for learning." This means that they might not be as sharp or as quick to learn as puppies. However, this doesn't mean that your dog is stupid or slow. Think about it; elderly people are still able to learn new things, such as how to use simple technological devices, the names of new people, and so on. Just as senior humans can learn new information and skills, senior dogs have this ability as well. To help improve the chances of your senior dog training's success, here are some tips to keep in mind:

- **Make some modifications to your dog's diet as needed**

    When it comes to your dog's diet, the best person to talk to about making modifications is a vet. But if you want to ensure the health of your canine as they approach the twilight years, you may want to decrease their protein and fat intake. Also, you can ask your vet about giving dietary supplements to your dog, such as chondroitin and glucosamine. After making any changes, no matter how small, observe your dog carefully to see how these changes affect their health.

- **Keep track of the treats you feed your dog**

  We have already mentioned how reward-based training works well for senior dogs. This means that you will be giving your dogs high-value treats while training. It's important to keep track of how many and how often you give treats to your dog though. That way, you can adjust their diet accordingly to avoid overfeeding.

- **Avoid repetitive physical activities or tasks**

  Keep in mind that senior dogs have older joints—and these don't do well with physical activities that are repeated again and again. This doesn't mean that you shouldn't include any physical tasks or activities in your training. Just make sure that you don't repeat these activities several times without giving your dog a break. This might cause your dog to feel pain, which, in turn, might discourage them from training.

- **Consider the temperature of the environment**

  Senior dogs have a higher sensitivity to extreme temperatures compared to younger ones. Observe your dog for any signs that they are feeling too cold or too hot while training. If you notice any discomfort, take a break and resume your training session when the temperature is more tolerable.

- **Use different kinds of signals while training**

  Apart from verbal commands, try to think of hand signals to pair with them. This helps reinforce your training, thus allowing your dog to learn faster. Hand signals are especially important if your dog is hard of hearing or has already gone deaf. But if your dog has already lost their sense of sight, verbal cues are all you need. Even blind dogs will benefit from dog training, so don't deprive them of it!

Book 1 - The Complete Guide To Dog Training

- **When your dog is tired, take a break**

    While training, keep an eye on your dog to spot any signs that they are tired, uncomfortable, in pain, or having breathing difficulties. If you observe any of these signs, take a break. Allow your dog to rest where they want to. After some time, you can continue your training session by picking up where you left off.

- **Don't force your dog to do things they aren't able to do**

    Dog training requires a lot of observation and planning. You need to know what your dog is capable of and what they aren't able to do anymore. Plan your training sessions accordingly. If you are trying to teach a command or trick, and your dog doesn't follow you even when you try to entice them with treats, don't force them. Make a note of this behavior and try again next time. If they still refuse, remove it from your plan.

- **As much as possible, have your training sessions in places with soft surfaces**

    This is especially important when you're teaching commands like "Stay," "Sit," "Down," and others that require your dog to get down on the floor. In such cases, prepare a mat or bed for your dog to train on. This will make it easier for your dog to respond or follow your commands, as they will be more comfortable.

- **Include stretches and other gentle exercises in your training routine**

    Gentle exercises such as stretching will help maintain your dog's flexibility, thus allowing them to do more. If needed, speak to your vet about what types of exercises are best for your dog and how much they should exercise each day. That way, you won't end up straining or overexerting your dog.

As you can see, all of these tips focus more on keeping your dog happy and comfortable while training. As with dogs of other ages, it's important to maintain positivity while training to make it worth your dog's while.

Chapter 7: Training Different Dog Breeds

When it comes to learning, does the breed of a dog matter?

This is a very common question posed by dog owners when they consider training their own dogs at home. If you pose this question to dog experts and professional trainers, they might tell you that a dog's breed doesn't matter when it comes to dog training. No matter what breed your dog is, you will be able to train them. Of course, this is true.

But as with other dog parents all over the world, you may be more interested in the level of difficulty when it comes to dog training—and that's why you would be asking the question. While all dogs can be trained, there are certain dog breeds that are considered easier to train and those that are considered difficult to train. But the important thing to keep in mind is that positive reinforcement training methods work for any kind of dog breed, even the more challenging ones. As long as you stick with your dog's training, you're sure to find success.

Dog Breeds and Behaviors

The main reason the ease of dog training may vary between dog breeds is that each breed has its own personality. Think about it; if you have to choose between a sweet-tempered dog breed who loves to learn or a stubborn dog who rarely listens to you, which of these breeds do you think would be more challenging for you to train? And if you're a beginner who doesn't want to be challenged, which breed would you choose?

Knowing that dog breed matters when it comes to the ease of dog training, you should carefully consider what breed of dog you should make a part of your life. Different dog breeds have their own distinctive personalities, which, fortunately, you can observe early on. In terms of personality, dogs can be grouped roughly according to the work they're supposed to do, and this can be a predictor of a dog's temperament when they grow up. Here are some examples of types of dogs and their common personalities:

- **Chasing dogs**

    In general, these dogs are extremely lively, feisty, and active, whether they are puppies or adults. They are also quite fast, which is why they are great at chasing.

- **Guarding dogs**

    These dogs have very strong protective instincts, which is why they are bred for the purpose of guarding flocks. They also have a very strong sense of loyalty.

- **Herding or working dogs**

    These dogs have dispositions that are almost business-like. They have a tendency to think about the situations in front of them before doing their tasks.

- **Hounds**

    These dogs have a tendency to be more independent and aloof. They aren't that interested in socialization with humans, and they have an inclination to do things on their own.

- **Sporting dogs**

    These dogs are adventurous and they love to follow their noses. Also, they always respond enthusiastically when their owners call their names.

These are just some examples of groups of dogs and the possible personalities they may have. Of course, there will always be exceptions to the rule. But more often than not, you can predict the behaviors of your dog based on these groups. It's also important to remember that all dog breeds have their own unique traits, thus making them appealing to different types of people.

Bringing a dog into your life is an important decision to make—and once you make that decision, remember that training your dog is your responsibility. The dog you bring home with you, regardless of breed, will rely on you for their care, health, and welfare. Caring for and training your dog is a lifelong responsibility. Therefore, you should take serious consideration as to whether your lifestyle allows for this kind of responsibility.

After making the decision to bring a dog home, it's time to think about the breed. Since there are some breeds that are easy to train and others that you might find challenging, knowing their breed's general personality can help you out immensely. So, llet's go through some of the most common dog breeds that are easiest and most challenging to train.

The Easiest Dog Breeds to Train

While training a dog is your responsibility as a dog owner, dog training doesn't have to be a difficult thing. Throughout the different chapters of this book, you have already learned a lot of practical and helpful tips for training dogs of all ages. All of this information can help you come up with an effective training plan and schedule to help your canine become healthy, happy, and well-rounded. If you're still thinking about the breed of dog to adopt, you'll be happy to know that there are certain breeds that are easier to train than others. As a beginner, training one of these dogs could potentially make your life a lot easier.

- **Border Collie**

Whether you already have a lot of dog training experience or this is your first time trying to train a dog, consider adopting a border collie. Dogs of this breed usually come with a lot of energy and always want to please their master. This means that such a dog will always be enthusiastic when it comes to training. However, because of their excessive energy, you should try to match this by coming up with fun, interesting, and stimulating activities for training and do these every day!

Fig. 13: Border Collie. From Unsplash, by Echo Grid, 2017, https://unsplash.com/photos/coheb5v8coM/Copyright 2017 by Echo Grid/Unsplash.

- **Border Terrier**

Most dog experts and trainers consider this dog breed to be highly trainable. Border terriers are fairly laid-back, but they do enjoy training and performing different activities. They are affectionate, trainable, and good-tempered too. As with border collies, this dog breed loves pleasing their master. You can use this to your advantage as you shower your dog with a lot of affection to keep them motivated while training.

- **Boxer**

This breed has an even temper and is very intelligen,t making it easy for them to learn new commands. Boxers are active dogs who immensely enjoy mental and physical challenges. They are playful, upbeat, patient, and have a protective nature, making them an excellent breed for families.

- **Doberman Pinscher**

While this dog breed is relatively easy to train, doberman pinschers are recommended for dog owners with more experience. As a beginner, if you are confident that you can provide consistent leadership and training to this dog breed, you can opt to take one home. However, don't let them get lonely or bored as this might cause them to become aggressive or destructive.

- **German Shepherd**

This breed of dog is always ready to work and eager to please. German shepherds are active, so they need a lot of mental stimulation and physical exercise. They are also well-known for their courage and loyalty. Furthermore, this dog breed also has the ability to learn commands and retain the information they learn for a long time.

- **Golden Retriever**

In particular, this dog breed makes an excellent companion for those who are new at owning dogs as they love pleasing their masters. Golden retrievers approach life playfully and joyfully, and they maintain this wonderful disposition longer than other dog breeds. One great thing about this dog breed in terms of training is that they learn through trial and error and have a unique ability to solve problems.

- **Labrador Retriever**

If you're looking for a popular dog breed that's easy to train, then this is the one for you. It's easy to train labrador retrievers whether you want to have one as a working or a family dog. This dog breed also socializes well with humans as well as with other canines. They are non-aggressive by nature, eager to please, and bursting with energy too.

- **Miniature Schnauzer**

It's easy for this dog breed to learn new tricks and commands. However, this is a very high-energy dog breed that you should keep occupied frequently. Miniature schnauzers crave companionship with humans. Combine this with the intelligence of the breed, and this makes training easy even for beginner dog trainers. This breed is also spunky, alert, and follows commands well.

- **Pembroke Welsh Corgi**

As another active dog who loves training, this breed loves their human companions and respond to training very well. Apart from training at home, Pembroke Welsh corgis would benefit a lot from enrolling them in obedience classes too. The more you train this dog breed, the more you will see them improve. For the rest of your dog's life, you will enjoy the company of a well-behaved and lovable canine.

- **Poodle**

Although there are different poodle varieties, all of them are eager to please, incredibly intelligent, and they learn quickly. This dog breed appears on different lists of the easiest dog breeds to handle while training. This is an even-tempered dog that aims to please their masters.

- **Rottweiler**

    This is another dog breed that's easy to train but will keep you busy too. Rottweilers have endurance, intelligence, and a willingness to handle responsibilities. This is why you will see a lot of rottweilers working as therapy dogs, herders, police dogs, and more. Just make sure to socialize this dog breed early on as they have a natural tendency to be territorial.

- **Shetland Sheepdog**

    Finally, this dog breed is very active and performs very well in different kinds of dog sports. Shetland sheepdogs are very easy to train, they're intelligent, and they will always want to please you.

The Most Challenging Dog Breeds to Train

As it is, dog training is already a challenging task, especially for beginners. Challenging but not impossible. While there are dog breeds that are easy to train, there are also those on the opposite side of the spectrum. If you feel intimidated with the task of training a dog and still haven't made a decision about which dog breed to choose, then you should re-think selecting one of these breeds.

- **Afghan Hound**

While this breed is lovable and faithful, you may experience a lot of challenges when it comes to training, mainly because this breed is extremely independent. Afghan hounds are independent thinkers and tend to be aloof, thus making training quite difficult.

Fig. 14: Afghan Hound. From Unsplash, by Arve Kern, 2017, https://unsplash.com/photos/35p5-Z3Zvzo/Copyright 2017 by Arve Kern/Unsplash.

- **American Pit Bull Terrier**

This breed is considered a fighting dog, and most of them are bred for this exact purpose. But when it comes to training, you might discover that American pit bull terriers aren't that easy to handle. Friendly, loyal, and loving as they are, this dog breed also tends to be very temperamental.

- **Basenji**

For this dog breed, the first thing you need to do is master crate-training as they tend to be quite stubborn. Basenjis are aloof, independent, and they are too intelligent for their own good. While they may learn the commands and tricks you are trying to teach them, the issue is whether they will perform these commands when asked.

- **Basset Hound**

What makes this dog breed difficult to train is their notoriously stubborn nature. It's especially difficult if you are a beginner at dog training. In particular, house-training basset hounds is extremely difficult. If you still choose to train this dog (such as if you already have one at home), then you need to have extra patience to stick with training.

- **Beagle**

As adorable and lovable as this dog breed is, you may experience a lot of frustrations trying to train them. This is because beagles are naturally naughty, thus making training a challenge—even for experienced dog owners.

- **Bulldog**

This is one dog breed that has become quite popular in recent years. However, as lovable as they are, bulldogs can be very stubborn too. This dog breed may look easy to handle, but they often won't care enough to listen while you are trying to train them.

- **Chihuahua**

This dog breed is extremely loyal, and they make incredible lapdogs. But they can also be very protective, fierce, and stubborn. As challenging as chihuahuas are to train, this is an essential part of raising this breed, especially in the area of socialization or they might grow up to be aggressive, overprotective, and might end up attacking other people.

- **Chow Chow**

Raising this dog breed to become well-behaved is an enormous task for any dog owner. Chow chows tend to be stubborn, dominant, and have an unpredictable temperament. They can sometimes be aggressive as well, especially when not trained properly.

- **Pit Bull**

This dog breed is powerful, strong, and originally bred as fighting dogs. Despite their tough reputation, pit bulls are actually eager to please and very friendly. While they respond well to training, what makes this dog breed difficult is that they get bored very easily. This means that you should stick to a very strict training plan and schedule unless you want your dog to chew, dig, or even become aggressive with other people, dogs, and pets.

- **Pomeranian**

This fluffy, adorable dog breed comes with a huge personality—and a reputation for being very difficult to train. Without proper socialization early in their life, pomeranians might become either too shy or too aggressive. House-training is also a challenge for this breed because they are very small. Also, this dog breed barks a lot, so training them to be quiet is essential.

- **Pug**

This dog breed is playful, loyal, smart, and sturdy; however, they can be a challenge to train. Pugs get bored easily, so you need to provide constant stimulation. Also, you need to stick with a very strict training routine to ensure that this dog will achieve the training goals you have set. In particular, this dog breed is very difficult to house-train.

- **Siberian Husky**

As beautiful as this dog breed is, raising them comes with a lot of challenges. For one, Siberian huskies are extremely active, especially while young. This means that you need to come up with a very dedicated training routine and stick with it. When they get bored, they tend to act out, which means that you should be very consistent when it comes to training them.

General Tips for Training Different Dog Breeds

Dog owners may often end up expecting too much from their canines while training. But when you think about it, expecting your canine to learn basic commands, household rules, and tricks is actually an enormous task. Just as you would teach a baby to crawl, walk, talk, and learn everything they need to survive in the world, training a dog requires a lot of time, patience, effort, and planning. Now that you know that there are breeds that are easier and harder to train, you can take your dog's breed into consideration when planning and scheduling their training. Even though some dogs are harder to train than others, it is entirely possible to train all dog breeds. No matter what breed of dog you own, here are some general tips for you to keep in mind while training:

- **Go slow, especially at the beginning**

  Even before you start training your dog, make sure that you have prepared an environment where you can train your canine without distractions. Such an environment allows your dog to concentrate fully when you are trying to introduce commands, tricks, and other information. Then, start off with one command at a time. Doing this makes it easier for your dog to learn what they need to and get rewarded for it. The more you reward your canine, the more they will start looking forward to training sessions and to learning new things.

- **Control your dog' training environment**

  Apart from preparing a quiet environment for your dog without distractions, there are other steps you can take to keep your canine focused on the task. Before bringing your dog inside the training environment, make sure that you have picked up any objects that might distract your dog, such as toys, shoes, and such. If you plan to train your dog outdoors, make sure to leash your canine first or use a longline. Also, train your canine within a fenced area.

- **Determine what motivates your dog**

All dogs are different, and this means that they will be motivated by different things. The great thing about training your dog is that you will learn more about them as time goes by. Over time, you will come to realize what truly motivates your dog and what doesn't. If you want to increase the likelihood of training success, you need to find the rewards that will encourage your dog to continue obeying you and following your commands. This is why it's more recommended to try out different treats and vary the types of treats you give. This keeps your dog interested, and it helps maintain the momentum of your training sessions.

- **Don't just focus on treats all the time**

When it comes to positive reinforcement, you don't have to use treats alone. There are so many other ways you can repay your canine when they do something right while training. Some examples of non-treat rewards are praise, petting, showing affection, and giving toys. Basically, anything that makes your canine happy would serve as an excellent training reward. If you recall, for a lot of the dog breeds that are challenging to train, one common trait they share is that they love pleasing their owners. So, if you don't think treats are working as effectively as you hoped, then you can try other types of rewards.

- **Be patient**

This is one tip that has appeared over and over again in the different chapters because it is just so important. No matter what breed of dog you have, no matter what size your dog is, and no matter how old your dog is, patience is essential if you want your training to succeed. Give your dog enough time to learn what you want them to. As long as you are consistent and you keep using positive reinforcement, you're sure to see the results of your training when the right time comes.

- **Tell your dog what you want them to do**

It's important to remain firm with your dog. There is no point in teaching your dog a specific set of commands on time then changing those commands after a few months. Once you have created the rules and routine you want your canine to follow, stick with those. Tell your dog exactly what you want them to do and keep practicing until your dog gains mastery. This is the most effective way your dog will learn, and it's also the easiest way for your dog to pick up the things you are teaching them to do.

- **Be careful with how you practice your commands**

An effective way to reinforce the commands you teach your dogs is practicing those commands as frequently as possible and making them part of your dog's routine. However, you should be careful when practicing those commands. For instance, don't use them only when you want your dog to do something they don't like. If your dog dislikes going into their crate (when you're not yet done with crate-training), don't just use the command "Come" when you want them to go inside their crate. Practice using the command in other ways too. If you only use the command for things your dog doesn't like to do, they might end up disliking the command as well. Mix things up and make sure that you use the commands you teach your dogs both in positive ways and for disciplinary purposes.

- **Make training part of your dog's life**

    We have mentioned how dog training is a lifelong process. Just because your dog has already learned the basic commands plus a few tricks, this doesn't mean that you are done with dog training. Keep in mind that training will provide your canine with the mental and physical stimulation they need to stay healthy. Apart from this, training also gives you an opportunity to bond with your dog each day. While you may reduce the frequency of your dog's training sessions over time, make sure that training becomes a permanent part of your dog's life.

Chapter 8: Common Behavioral Issues in Dogs and How to Deal with Them

Dog training doesn't involve teaching your canine basic commands and tricks only. You can also use it to help "fix" behavioral issues that your dog may have developed or learned throughout their life. Of course, we all want our canines to be well-behaved and happy. This makes it easier for us to live in harmony with each other. As a dog owner, you may experience a number of common behavioral issues at one point or another. In such a case, you must learn how to deal with these issues, and the best way to get rid of them is through training.

Training Hyperactive Dogs to Calm Down

When you have a hyperactive canine, using positive reinforcement training is key. High-energy dogs are very common; however, you shouldn't allow your dog to continue being hyperactive all the time as this will make it very difficult to live with. Also, if left unchecked, a hyperactive dog might learn other destructive behaviors that might be even more difficult to unlearn. If you have a hyperactive dog, here are some tips to help them calm down:

- Manage the excess energy your dog has by making sure that you include a lot of exercise into their daily routine, whether during training or during your free time. Until your dog is fully trained, make sure to leash your dog while doing these physical activities. If you have a fenced yard, you can allow your canine to run around outside for a few minutes throughout the day.

- Another way to help your dog release all that pent-up energy is by taking them to the dog park. This allows your dog to play with other dogs while getting the exercise they need and learning how to interact with other canines.

- Playing games with your canine is an excellent activity too. There are different types of games you can play with your dog both inside and outside of your home. Some great examples of games to play with your canine are:

    - **Find the Treat**

        This is an excellent game to play with dogs who love to explore their environment using their noses. Introduce this game by holding a treat in one hand, tossing it to your other hand, and giving the command, "Find it!" Of course, your dog will find the treat easily. Do this step around 6 to 7 times for your dog to understand the command.

        Then, ask your dog to "Stay" while you walk away from them. Make sure your dog is watching you as you place the treat on the ground. Go back to your dog, pause for a moment, and give the command, "Find it!" Again, do this step around 6 to 7 times for your dog to understand the command.

        Now that your dog understands the game, you can start hiding the treat in different places. Start by hiding the treat in easy-to-find places to give your dog more opportunities to practice. The more you practice, the more difficult locations you can use as your hiding spots.

    - **Hide and Seek**

        This is another great game to play with your dog. It's similar to the first game but this time, you would tell your dog to "Find me!" instead of "Find it." Start by hiding in easy locations and giving your dog a treat when they are able to find you.

- Finally, you can also use positive reinforcement training to calm your dog down. The more you use training to calm your dog down, the more they will learn that being too active all the time isn't okay. There are different types of positive reinforcement training methods you can use for this issue, including:
    - Use the command "Sit" as your dog's default behavior. When you feel like your dog is starting to become too energetic, give the command and allow your canine to relax. If your dog is able to sit for an acceptable amount of time, reward them with a treat.
    - If your dog is being hyperactive because they are excited, give the command to "Wait" and ignore your dog until they calm down. Once calm, give a treat and proceed with the activity you were going to do in the first place.
    - Use clicker training to help your dog understand that calm behaviors deserve clicks and rewards!

Play Biting, Mouthing, and Nipping

While playfully biting, mouthing, or nipping is normal for dogs, especially puppies, allowing these behaviors to continue into adulthood isn't a good thing. While your puppy might not hurt you when they nip or bite you, these behaviors will become a problem very quickly. If you want to avoid injuries and other issues that might result from these behaviors, here are some tips on how to deal with them:

- Whenever your dog tries to bite, mouth, or nip at you, use a toy as a distraction. Push the toy at the mouth of your dog and try to start a short tug-of-war game. You can give the command

"No" when your dog is biting at you, then another command like "Get toy" as you push the toy at their mouth.
- Provide your canine with a wide range of interesting and fun chew toys to keep them distracted. The more toys they have, the less interested your dog will be in biting, mouthing, or nipping at you.
- You can also try to use treats to distract your canine when they start biting, nipping, or mouthing at your hands. Don't use treats too frequently, however, as your dog might learn that these behaviors are good because they get treats after doing them.
- Try not to panic if your dog bites, nips, or mouths at you even after you give them the toy. But if your dog continues the bad behaviors after you gave them the toy, you should stop the behavior. Again, give the command, "No," this time more firmly, to emphasize that these behaviors aren't okay. If you have to, walk away from your dog so that they learn that when they do these things, you won't play with them.
- If your dog nips at your heels frequently, distract them with a tug toy. Stop moving or walking, then wave the toy in front of your dog's face enticingly. If you aren't carrying a toy, you can simply stop and wait until your dog stops mouthing or nipping at your heels.
- You can also train your dog to perform bite inhibition, the ability to control their mouthing force. Developing this ability in your dog allows them to recognize that the skin of humans is more sensitive. To train your dog to be more gentle, allow them to mouth on your hands. But when your dog bites you hard, give out a yelp immediately, then allow your hand to go limp. This comes as a surprise to your dog, which may cause them to stop. If your dog stops (or if they start licking you), give them a reward, then continue with your game. Repeat these steps 3 times before you take a break. Keep practicing until your dog learns how to be more gentle while playing with you.
- Dogs can also learn bite inhibition by playing with other dogs.

In fact, this is where they learn more naturally. Therefore, if you think these behaviors are starting to become a problem, try to increase your dog's social interaction and playtime with other dogs.

- Focus on playing games with your dog that don't involve contact with your hands. This helps satisfy your dog's urge to mouth or bite without hurting you. However, this doesn't mean that you shouldn't play with your dog in general. Play is an important factor that strengthens your bond with your canine. Just choose your games wisely and don't encourage inappropriate behaviors when they happen while you are playing.
- Finally, never use physical punishment, such as hitting or slapping your dog. This might cause your dog to bite you harder, and it might teach them to become aggressive. As with other issues, focus on positive reinforcement when trying to eliminate these behaviors.

Dealing with Aggression

This is one of the more troublesome issues to deal with. After all, young or old, you wouldn't want to share your home with an aggressive dog, especially if you live with other family members, children, or pets. Aggressive dogs snarl, bare their teeth, pull back their ears, and show other scary behaviors. Usually, aggressive canines exhibit such behaviors when they are feeling territorial. If you start with a puppy, then you can discourage aggression early on. Even if you have a dog breed that has a territorial nature, if you shower them with love, affection, and positive reinforcement, you might not have to deal with this issue when they grow up.

Fig. 15: Aggressive Dog. From Pixabay, by Christels, 2017, https://pixabay.com/photos/dog-bad-dangerous-german-shepherd-2141358/Copyright 2017 by Christels/Pixabay.

However, if you have adopted a young, adult, or senior dog and discover that they display aggressive behaviors or tendencies, you must work on eliminating these right away. There are many reasons why dogs behave aggressively. To deal with the issue, you should first determine the cause of their aggression. They may react aggressively to something you do, because they are protecting something, because they think you might hurt them, or because they are in pain.

Aggressive dogs are very intimidating, especially when you don't know the reason for their aggression. The key is to remain calm and to never force your dog to do something they obviously don't want to do. It's important to observe your canine carefully, have patience, and approach the situation positively. Here are some tips to help:

- If your dog starts snarling when you get close to them while they are eating, don't approach. This is a natural instinctive reaction. For this, you can casually walk by your dog while they are eating. Don't approach them directly and don't attempt to take away their bowl. Just get your dog used to your presence so that they learn that you aren't a threat even if you stand or walk close to them. You can follow the same approach when your dog is acting aggressively because they are trying to protect something.
- If your dog reacts aggressively when another dog is close, distract them by calling out their name. If your dog looks at you and relaxes, give them a treat. If not, call out their name again while gently tugging on their leash. When rewarding your dog, show enthusiasm so that they know that looking at you instead of at the other dog is a good thing.
- If your dog is acting aggressively because they think you might hurt them, don't force the situation. This may be a common issue for dogs who have been abused in the past. Of course, your canine won't be acting this way all the time. Whenever your dog is acting calmly toward you, shower them with love, affection, and enthusiasm. Over time, your dog will learn that you aren't a threat.
- You will notice right away if your dog is acting aggressively because they are in pain. Your dog might flinch whenever they move. In such a case, take your dog to the vet and have them checked. The vet can give your dog medications to ease the pain and, hopefully, eliminate the aggressive behaviors too.

When it comes to aggression, the bottom line is to never approach or force a dog who is showing such behaviors. Determine the cause of the behavior, and this will help you come up with a plan to eliminate it.

Chewing, Digging, and Other Destructive Behaviors

Chewing is a natural and normal activity that all canines do. It's also important to help alleviate their anxiety and maintain the health of their teeth. Dogs explore their world through their noses and mouths, and chewing is a method of exploration for them. Unfortunately, allowing your dog to chew everything in sight can be very destructive. To help eliminate this behavior, you can:

- Provide your dog with different kinds of chew toys. Make sure that these are safe, interesting, and fun to chew on.
- Observe your dog as they explore their surroundings. If they sink their teeth into something they're not supposed to, give the command "No" and replace that item with one of their chew toys.
- Make sure that you don't leave your things lying around the house, especially items that are easy to chew and destroy.

Digging is another destructive behavior that you should help your dog manage. If you have a dog who loves to dig, you should first try to determine the reason for this destructive behavior. If you notice that your dog digs when they are bored, you can play with them more or give them more toys to keep busy throughout the day. In some cases, dogs who are left outside feel vulnerable, and so, they start digging. If you think this is your dog's reason for tearing up your yard, don't leave your canine outdoors for too long. But if your dog is trying to escape, you may need to extend your fence deep into the ground so that they won't be able to. Finally, if you think your dog simply likes digging, it's best to set an area where it's okay for them to dig. Praise your dog for digging in the right spot and reprimand your dog when they dig anywhere else.

There are other types of destructive behaviors that dogs may engage in, and it's important to help your dogs eliminate these behaviors. This process involves training, practice, and consistency. Here are other destructive behaviors that dogs commonly do and how you could get rid of them:

- **Stealing food** is both destructive and dangerous. For this behavior, you should teach your dog some self-control. Do this by teaching the "Wait" command and practicing it in different situations. The "Leave it" command works well here too. After your dog masters these commands, you can go further by teaching your dog what's theirs and what's yours. For this, you can place your food in front of your dog and give a command like "Not for you." If your dog tries to lunge at the food, stop them and repeat the command firmly. If your dog walks away or ignores the food you have placed in front of them, reward them with praise and a treat. Keep practicing this until you notice the behavior disappear.

- **Separation anxiety** is an issue that may cause your dog to start destroying things at home when you leave them alone. You know your dog suffers from this condition when they start becoming very anxious as you prepare to leave. Also, you may start observing the destructive behaviors within 15 minutes after you have left. When you are at home, your dog will constantly follow you and will always try to maintain contact with you as much as possible. To deal with this, you may have to perform behavior modification, desensitization exercises, and a lot of dedicated training sessions. You may want to seek professional help for your dog to overcome this issue.

Barking, Whining, and Other Noisy Behaviors

There's nothing more irritating than a dog who barks non-stop, especially at night. This noisy behavior is common in dogs of different breeds, and it can be quite frustrating to help your dog overcome it. But as with other behavioral issues, excessive barking can be eliminated with these proper training techniques:

Fig. 16: Barking Dog. From Pixabay, by dahancoo, 2017, https://pixabay.com/photos/doodle-barking-dog-woof-2965983/Copyright 2017 by dahancoo/Pixabay.

- If your dog barks non-stop at something, call their name and give the command "Come." This shows that you acknowledge the presence of the thing they are barking at, but you want them to stop barking. If your dog stops barking and comes to you, reward them with a treat.
- If your dog barks at every little thing they see outside, all you have to do is cover up your windows. For most dogs, this solves the barking problem.
- If your dog keeps barking because they are overstimulated, you may try giving them a time out. Do this if the first two tips don't work. Bring your dog to a quiet place and leave them there for some time until they calm down.
- Teach the commands "Quiet" and "Speak" to help your dog learn that there are times when it's appropriate for them to bark and times when they should be quiet. Simply give the command and immediately give a reward when your dog performs the correct action.
- If nothing else works, you may consider using a bark collar

(preferably one with a citronella spray, not one that will shock your canine). However, you shouldn't keep this collar on your dog all day—only when they start their barking frenzies.

If your dog whines a lot, the most common reason for this is that they are feeling anxious. Try calming your dog down by showing them some love and affection. Then, you can try the "Quiet" command as you are reassuring your dog. If your dog is whining just to get your attention (for instance, while you are crate-training), ignore the behavior. This will help your dog understand that the behavior won't work as they intend. Once they stop, reward your dog with a treat. When it comes to noisy behaviors, finding out the cause is essential. Dogs will always have a reason for making noise. Once you discover the reason, you can work on fixing the issue!

Chapter 9: Rewards and Punishments

The choices that dogs make in their lives (and the behaviors that result from these choices) are generally based on whether these choices result in a reward or a punishment. With this in mind, you can say that training involves the manipulation of rewards and punishments in a way that will cause your dog to always make the right choices. If you want dog training to work for you, it's important to learn as much as you can about your dog in order to find out which rewards will truly motivate them.

Back in the 1950s, a man named B. F. Skinner, a behavioral scientist, came up with several principles that apply to any living thing that possesses a central nervous system. One of these principles is that living things have a high likelihood of repeating behaviors that are rewarding to them. Also, they have a low likelihood of repeating behaviors that result in punishment or anything unpleasant. Of course, the application of this principle is quite obvious when it comes to dog training. When you give rewards to your dog for good behaviors, they will most likely repeat them. By contrast, when you punish, ignore, or discipline your dog for doing bad behaviors, they will most likely avoid doing these behaviors again in the future.

Although the definition and explanation of this principle are very clear, when applied to dog training, it isn't that simple. The main reason for this is that punishments and rewards may vary from one dog to another. For instance, one dog may find treats to be highly motivational while another dog prefers affectionate gestures from their owner. On the other side of the spectrum, one dog may learn quickly when you ignore their bad behaviors while another dog may need firmer or stricter methods. To put this in the simplest words, a reward is something your dog likes, and punishment is something your dog doesn't like.

While training dogs, most dog owners use dog treats as rewards. Treats are very effective, as most dogs absolutely love them. When you need to teach tougher commands or tricks, you may opt for high-value treats that are more delectable and appealing to your canine companion. If you have an affectionate dog, you may also use praise, petting, and other affectionate gestures to motivate your dog while training. When it comes to finding the perfect reward, it's up to you as the owner to find out which reward will work best—and you can determine this by learning more about your canine companion.

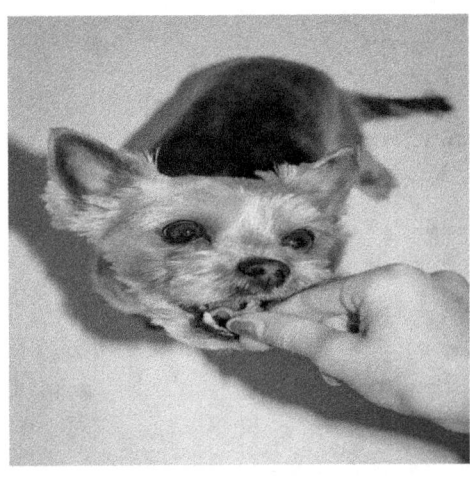

Fig. 17: Treats. From Pixabay, by Pezibear, 2019, https://pixabay.com/photos/dog-yorki-pet-small-sweet-cute-4047041/Copyright 2019 by Pezibear/Pixabay.

Punishments, on the other hand, are even trickier. Since positive reinforcement training methods are more recommended than those that involve harsh punishments, how can you punish your canine to teach them to avoid bad behaviors? Well, one effective way to eliminate bad behaviors is to ignore them. Of course, this may not work for all kinds of situations or behaviors. Punishments or discipline should still be part of your dog training though. Later in this chapter, we will go through the most effective ways of disciplining your dog so that you don't have to deviate from positive reinforcement methods of training your furry friend.

Reward-Based Training

When you hear the term reward-based training, your mind probably immediately goes to dog treats, as does ours. When it comes to this type of training, dog treats are the star—and they can be extremely effective for teaching commands and other obedience training concepts. This method of training harnesses the power of a very primal thing—food—and it makes learning easier for dogs too.

While most dog owners opt for dog biscuits, these might not work well for all dogs. Some dogs may find these boring, and others might not like the taste. If your dog is a fan of dog biscuits then, by all means, use them! But if your dog isn't interested in these reward, you may have to find other dog treats that will awaken their desire to learn. Sadly, choosing the wrong rewards can have an adverse effect on your training. For instance, if you taught your dog a command, they were able to do it on their first try, and you rewarded them with a treat they don't like, this will decrease the likelihood that your dog will repeat the behavior again in the future.

There is no standard formula for reward-based training. It is a trial and error process wherein you will have to observe your dog to see what they like, what they don't like, and what catches their interest the most. You can come up with a "hierarchy of rewards" for your dog and use this list while training. The more difficult the task is, the higher their reward must be. For instance, if your dog likes biscuits but loves turkey slices, you can use the biscuits to reinforce practices and give turkey slices when you introduce new commands. It's all about matching the reward with the task at hand.

The timing of giving your dog a reward also matters immensely. Right after your dog does something good or right, make sure to give the reward right away. This helps your dog understand that they are getting the reward precisely because they performed the action that you were asking for. When you give the reward even a few seconds after your dog performed the action, they might not understand why they received it in the first place. Timing can either reinforce behaviors or end up confusing your dog. Therefore, you must make sure that your timing is perfect! Here are more tips to keep in mind when it comes to reward-based training:

- The placement of the rewards you give matters too. For instance, if you taught your dog to "Sit," give the treat right after they sit and while they are still in the sitting position. Think about it; if you praise your dog when they sit but give the reward when they stand up, what do you think your dog will learn from this? While timing the reward, make sure your placement is on-point too.
- While training your canine, use light and small treats. These should make your dog feel good, but they shouldn't cause your dog to gain excess weight because you train them frequently. Keep track of the treats you give your dog and adjust their meals as needed.
- Use treats as rewards, not bribes. This means that you should give them after your dog does the command, not before.
- It's best to give your dog a reward when they are in a calm and submissive state, not when they are hyperactive or excited. Otherwise, you might end up reinforcing the wrong behavior.
- When you are trying to teach your dog a command that involves a lot of steps, you may give them a reward for each step they complete. In such a case, you may use simpler treats for the steps and a high-value treat when your dog reaches the end of the process.
- As you practice commands and behaviors over and over again, try eliminating the use of treats gradually. That way, your dog learns to follow your commands even without treats. When your dog is able to do the commands without the need for treats, you know that they are approaching mastery.

Positive Reinforcement Training

Positive reinforcement training is the easiest and most effective way to train your dog. With this method, you give your dog a reward when they perform good behavior in order to reinforce that behavior. This training method is very powerful, but it's more than just providing praises and rewards. To ensure that you get the most out of this method, you must find the perfect rewards and praise for it. Using these rewards and positive reinforcements will make your training interesting and fun for you and your furry friend. Here are some tips for you to improve your positive reinforcement dog training:

- **Involve everyone in your dog's training**

    This tip applies if you live in a household with other people, such as the members of your family. After reading this book and learning everything you can about dog training, you can share this information with the other members of your household. That way, they can take charge of training your dog once in a while, especially when you're not around. Involving everyone also means that they take part in reinforcing your training by ensuring that your dog's routine is followed each day, every day. Just make sure that they understand what positive reinforcement means and how to implement it.

- **Vary the rewards you give**

    When it comes to positive reinforcement, treats aren't the only rewards you can give. If you give your dog the same type of reward or the same type of treats over and over again, they will probably get bored. But if you mix things up once in a while and shift between new rewards, rewards your dog really loves, and some "ordinary" rewards, this will keep your dog wanting more.

Book 1 - The Complete Guide To Dog Training

- **Establish communication with your canine**

    By nature, positive reinforcement methods help you establish communication with your dog. Since you will be giving your canine something positive in return for doing something good, your canine will soon come to understand which behaviors they should do and which aren't appropriate. Make sure to communicate clearly with your canine. And when your dog is doing something that isn't right, be both firm and clear about this as well. This will help give your dog a clearer distinction between what is right and what is wrong.

- **Include games and play in your training sessions**

    Games and play can be fun and effective rewards for your canine. But don't use them in the middle of your training session as this might leave your dog confused. Instead, when your training session is coming to an end, you can ask your dog to do something you know they're already familiar with. When your dog performs the action correctly, you can start a game or initiate play as their reward. That way, you can extend the bonding time you have with your dog, end the training session on a positive note, and make your dog appreciate you more.

- **Also, include outdoor activities in your training sessions**

    Dogs love going outside! So, once in a while, you can make your training sessions even more fun by having them outside. Of course, you should only do this when you are reinforcing concepts or commands that you have already introduced. However, it's not a good idea to train your dog outside when you are trying to introduce new concepts, commands, or information to your canine. With all the new stimuli in the great outdoors, your dog will have a very hard time focusing on you.

- **Make your training sessions short, sweet, and simple**

    This is especially important at the beginning of your training. When making your training sessions, don't try to have sessions that are half an hour long or more. Remember that dogs have shorter attention spans than us humans. Also, don't plan training sessions in which you will teach your dog 4 or more commands or tricks at a time. If you want to get the most out of each training session, keep them short, limit the information, and provide a lot of positive reinforcement.

- **Always correct your dog as needed**

    Finally, although positive reinforcement training focuses more on reinforcement, you shouldn't be afraid to correct your dog as needed. Yes, you should encourage and reinforce their behaviors using positive methods. But if your dog is doing something bad, dangerous, or inappropriate, you should correct the behavior too. Just make a clear distinction between your reinforcements and your corrections to ensure that your dog only learns the right things.

Should Punishments Be Part of Your Training?

In terms of dog training, punishment refers to something that happens around or to a dog that reduces the likelihood that they will repeat the behavior that preceded it. Punishment can come in the form of deliberate things, such as a stern warning, a physical hit, or something else. It can also come in the form of accidental things, including stepping on something sharp, or coming into contact with something hot. No matter what, as long as the dog experiences something painful or unpleasant, this can be considered as punishment.

Now that you know this definition of punishments in the context of dog training, should you include them as part of your training?

That depends.

Physical punishment such as hitting, beating, or spanking your dog has no place in positive reinforcement training. These are considered aversive techniques wherein a dog feels pain where they are hit. While using physical punishment may eliminate bad behaviors, this might also cause other adverse effects on your dog. For one, your dog might learn to become aggressive if you always hit them. Some dogs become fearful of their owners too. Also, when you hit your dog, you can forget about forming a strong and loving bond with them. Most likely, your dog will see you as nothing more than an authority figure they must obey unless they want to get hurt. Apart from the pain they feel, physical punishment may also increase the dog's stress significantly while lowering their quality of life. Why do this when there are better ways?

There is such a thing as "positive punishment" in which you would do something around or to your dog that they don't like. This is considered "positive" because it doesn't involve harshness and pain; you are simply "adding" something to the environment of your dog that they don't like. These things are called "aversives," and they come in different forms. Some of the more common aversives are rattle bottles, heeling sticks, spray collars, and so on. You may consider using these in your training as an alternative to negative punishment, which may harm your dog in different ways.

Effective Ways to Discipline Your Dog

Disciplining your dog is part of the training process. Without discipline, you might just end up with a spoiled adult dog who does whatever they want, whenever they want. And you don't want to have a dog like that. Although punishment is usually seen as a bad thing, it doesn't always have to be. Disciplining your dog by incorporating punishments will help your canine understand what they are allowed to do and what they aren't allowed to do. There are simple kinds of punishments that you can carry out that are both efficient and humane. Some examples of these are:

Fig. 18: Dog Discipline. From Unsplash, by Darinka Kievskaya, 2018, https://unsplash.com/photos/ff22 1Bu56mI/Copyright 2018 by Darinka Kievskaya/Unsplash.

- Taking your dog's toys away.
- Using a firm tone to stop your dog's behavior and pairing this with an equally firm restraining action.
- Ignoring your dog until they stop the behavior.

As you can see, these punishments give a clear message, but they won't make your dog feel wary, fearful, or angry at you. Consistently giving these punishments will help your dog see and understand that you won't tolerate the behavior they just did. When it comes to giving punishments while training, here are some things to keep in mind:

- When thinking of a punishment to give your dog, make sure that it matches the severity of the misbehavior for it to stick. For instance, if your dog is simply acting too excited, your punishment can be to ignore them. But if your dog is doing something dangerous or inappropriate, you should think of a more urgent and firm punishment to let your dog know that what they are doing isn't okay.

- Just as you give rewards, make sure to give your punishment immediately after your dog's behavior. This will help your dog understand that the behavior that you are punishing is the one they have just done.
- When dealing with an aggressive dog, be very careful in giving punishments. Never approach such a situation with heightened emotions. For instance, if your dog is growling at you or a member of your family, never lunge at your dog while giving a stern command. This might surprise your canine, causing them to lunge back at you or even bite you. Instead, remain calm while you are giving your stern command. Once your dog calms down, you can give them a treat and shower them with affection. Of course, you should also learn the cause of your dog's aggression so that you can help avoid the behavior entirely.
- Finally, after punishing your dog, make sure to teach a good behavior in replacement of the bad one. Distracting your dog by teaching them a good behavior may help your dog forget the bad behavior more easily. Doing this teaches your dog how to behave when faced with a specific situation. In some cases, when you don't teach a new behavior, your dog becomes confused. And when the same situation arises, your dog might just go back doing the behavior you punished them for.

Dog discipline involves training your dog to differentiate acceptable behaviors from unacceptable ones. Using positive punishment or disciplinary techniques is a lot more effective than trying to strike fear into your dog by hitting them or being very harsh with them. Here are a few more effective tips to help you discipline your dog without using harsh punishments:

- First, keep in mind that your dog's behavior isn't malicious or purposely done. More often than not, dogs are just bored, or they do things because these behaviors seem "normal" to them. If you catch your dog doing something wrong, it's up to you to educate them.

- You may want to think of a correcting cue to say when your dog is doing something wrong. The common ones are "No," "Think again," "Stop," and "Leave it," each of which convey a clear meaning. Train your dog to learn the meaning of this word or phrase by saying it each time your dog does something bad, then rewarding them if they stop.
- Try doing the "scruff-shake." For this, grab hold of the fur under your dog's neck or ears and give them a brisk but brief shake. Do this when you catch your dog in the act of doing something bad. This action comes as a surprise to your dog, thus stopping the behavior immediately.
- Also, try doing the "put-down" move. To do this, push your dog over gently but quickly onto their back or side then lean over them. This shows your dog that you're dominant, and they must follow you.
- Time-outs work well too, and you can do them in different ways. You can ignore your dog, stop interacting with them, stop playing with them, or place them in an isolated place for a couple of minutes. Do this right after you catch them doing something bad.

When it comes to doggy discipline and giving punishments, the key is to find the perfect balance between helping your dog learn that the behavior they are doing is unacceptable and considering your dog's feelings. You shouldn't be too strict nor should you be too much of a pushover. Remember, you are only doing this for the good of your dog. So, stick with your disciplinary methods to ensure that your canine will learn what's right and what's not.

Chapter 10: Advanced Training Tips and Tricks

Dog training is truly important. It's something that no dog owner should ever ignore! A lot of dog owners out there may feel complacent as long as their dogs aren't aggressive or don't misbehave all the time. But even if your dog is fairly well-behaved without training, you should still engage in dog training—and now you know why. Unless you are aiming to teach your dog complicated tricks or train them to do a specific task, dog training doesn't have to be that difficult. As long as you can teach your dog the basic commands and a few simple tricks, living with your dog will be a whole lot easier. Plus, you and your canine will be able to enjoy all the benefits dog training has to offer.

Once you have made the commitment to include dog training into your dog's daily life, you may soon discover how much fun this challenge is. The closer you get to your dog, the more you learn about them. And the more you see how your dog improves and progresses, the more motivated you may become to push yourself and your canine companion even further! If you have finished all the basics and want to encourage your dog to learn more, this chapter will be a huge help. As long as you make dog training a positive thing, your dog will surely love every minute of it. And since dog training is a lifelong process, learning these advanced tips, tricks, and strategies will help you out a lot.

Precision Dog Training

Have you ever heard of precision dog training? This is a more advanced type of dog training wherein you go beyond the basics to make sure that you always have your dog under control no matter what the situation may be. Although you may have no plans to have your dog compete in shows, precision dog training is something you can do to enrich your dog's life and improve their behaviors.

Fig. 19: Precision Training. From Pixabay, by MabelAmber, 2018, https://pixabay.com/photos/dog-animal-mammal-canine-pet-3194862/Copyright 2018 by MabelAmber/Pixabay.

By the name itself, you should know that this type of dog training is more strict and more precise. Only the most dedicated dog owners take their dog's training further than the basic commands. If you're one such dog owner, then you can also give precision dog training a try. After you have finished this basic training and are confident with the commands and skills your dog has learned, then you can move on to the more advanced stuff. Never rush basic training as this will be the foundation of other types of training you plan to do in the future. Therefore, you can only consider doing precision training and other advanced training methods when you have observed that your dog shows consistency in terms of following basic commands along with the routine and rules you have set.

As a dog owner, it is your choice whether you want to continue your dog's training or not. While this is highly recommended, it's understandable for some people to just stick with the basics. There is nothing wrong with this, of course, as long as you maintain your dog's training throughout their life. But if you're interested in precision dog

Book 1 - The Complete Guide To Dog Training

training, some of the most common things to teach your dog are:

1. **Off-Leash Training**

    While your dog follows all of your commands when you are in familiar environments or while your dog is on a leash, don't expect the same once you remove the leash and your dog realizes that they are free. Obviously, the difference between off-leash training and basic training is that the former doesn't involve a leash. For this type of training, you will re-train your dog with the basic commands and other tricks, only this time, you won't use a leash. As with basic training, start slow.

    Don't attempt to train your dog off-leash for the first time in a dog park or other equally busy environment. Start in distraction-free, enclosed environments, allowing your dog to explore without their leash. Then, teach the basic commands and keep practicing them each day. For this type of training, the most important command to help your dog master is "Come." That way, no matter how big the environment is or how far your dog goes, they will surely return to you when you give this command.

2. **Using Hand Signals**

    Another way to enhance your training and reinforce the commands you teach your dog is by teaching hand signals. This is especially beneficial if your dog's hearing starts to weaken as they age. Hand signals are also very effective when you are training your dog off-leash and from a distance. Hand signals are your secondary way to communicate with your dog.

    The choice of what signals to use is up to you. As long as your dog can differentiate between the signals you come up with, you don't have to worry. You can use the same signals as those used by professional dog trainers or you can come up with your own unique signals to indicate the commands. Keep practicing these signals both on and off-leash until your canine understands that they mean.

Tips for Training Working Dogs

Working or service dogs help people in their daily lives so that they don't have to depend heavily on others. In some cases, owners may want to train their dogs to help out at home and serve as working dogs even though they haven't been professionally trained. In the simplest terms, working dogs are trained to take specific actions as needed in order to help their masters. The dog is able to perform tasks according to what you have trained them to do.

Working dogs aren't required to be trained by professionals. This means that you can also train your dog at home according to your own needs and the tasks you want them to do. Before training, make sure that your dog meets shows the following characteristics:

- Is willing to please you
- Can reliably perform tasks over and over again
- Is able to remain calm no matter what surroundings they are in
- Is able to socialize in different environments and situations
- Has the ability to learn information and retain what they have learned
- Remains alert, though not reactive

If you want your dog to become a working dog, help them master the basics first, including house-training, socialization, and the basic commands. Also, make sure that your dog learns to focus solely on you no matter how many distractions there are in the environment. Then, you can start thinking about what tasks you want your dog to learn. Upon determining these tasks, you can start training your dog to take a specific action when you need assistance. To help you with training, here are some tips to keep in mind:

- Start your dog's task-specific training as early as possible.
- Build a genuine and loving relationship with your dog.
- Make sure that your dog is at the peak of health and receiving enough exercise time to remain strong.
- Groom and handle your dog with care each day.
- Enrich your dog's mental health by providing a lot of activities that involve nose work, mind puzzles, playtime, and socialization.
- Help your dog master obedience training before transforming them into a working dog.
- Allow your dog to have some fun too.

As with other types of dog training, this takes time. If you want to have a working dog, put in the time and effort to train them. Also, keep practicing the basics once in a while so that your dog never forgets them.

Training Small Dogs

Most of the time, smaller dogs are able to do things that big dogs aren't allowed to do, and this is mainly because of their diminutive sizes. However, this doesn't mean that if you have a small dog, you should allow them to do whatever they want at home. While you would use the same basic principles and techniques for training bigger dogs, there are some size-specific tips you may want to use to ensure that your tiny dog's training goes smoothly. Here they are:

- **Go down to your dog's level**

    Because of their size, small dogs may feel intimidated or threatened by how big you are. Instead of looming over your dog, go down to their level. This helps prevent your dog from becoming unresponsive, intimidated, or defensive while you are training.

- **Help your dog master the basic commands to ensure their safety**

   To ensure the safety of your small dog, the first commands you should teach them are "Sit," "Stay," and "Down." These will help your dog become more well-behaved, and they will also help prevent your canine from darting away from you, especially in unsafe environments.

- **Make sure your dog knows who is in charge**

   A lot of small dogs tend to feel like they are the rulers in their homes. If you allow this thinking to continue, you'll end up having a little diva who doesn't listen to you. From the get-go, make sure that your dog knows who is in charge—and that's you! Do this by remaining firm with your dog, especially when they are doing inappropriate or wrong things. The more your dog realizes that they can't get away with everything, the more they will come to respect you and see you as their master.

- **Try different training methods to see what works**

   Just because you have a small dog doesn't mean that they cannot learn as much as big dogs. As with any other type and size of dog, treat your miniature one as a blank slate. Try different things to see what works, what your dog excels at, and what you may need to change in your training plans.

Training Large Dogs

If personality is a big issue for smaller dogs, handling is the main issue for large ones. Imagine having to calm down an enormous dog when they are excited about something! When it comes to large dogs, obedience training becomes even more important as the risks that come with misbehaving increases significantly as the size of the dog increases.

Training is essential to ensure your safety, your dog's safety, and the safety of those around you. Here are some tips to keep in mind when training large dogs:

- **Start off with basic commands to help you control your dog**

    As with small dogs, there are basic commands that you may want to teach first in order to make it easier to live with, manage, and train. These basic commands include "Heel" (learning to walk on a leash), "Down" (not jumping up at you), and "Get off" (not climbing on the furniture).

- **Make eye contact as you give commands in a firm way**

    Performing these two actions while training your dog helps them learn that you are serious about your commands and will not back down. While maintaining eye contact, give your commands in a clear and loud voice. This will make an impression on your dog, even with their intimidating size.

- **While training your dog, make sure that there aren't any children or animals in the room**

    Training large dogs in quiet and calm environments is the way to go. Otherwise, your large dog will get distracted or excited with the presence of children, other dogs, or other types of pets. Also, keeping them away helps ensure their safety while your dog hasn't mastered their training yet.

More Advanced Training Tips and Tricks

When it comes to dog training, the possibilities are endless. There will always be more things to teach your dog. But this doesn't mean that you should rush your training sessions just that so you can continue teaching your dog more and more things. Remember that this is something you will be doing throughout your dog's life. So, take things slow, give your dog time to master commands and tricks, and keep these final tips in mind:

Fig. 20: Dog Tricks. From Pixabay, by Katrin B, 2015, https://pixabay.com/photos/dachshund-dog-school-dog-training-672780/Copyright 2015 by Katrin B/Pixabay.

- **If you notice that your dog has excess energy, use training to help burn this energy off**

    Allowing your dog to burn their excess energy in productive ways will help you avoid a lot of issues. Often, dogs who have too much energy end up being noisy, destructive, and frustrating. You can't blame your dog as they aren't in control of how much energy they have. Instead, help your dog overcome this by allowing them to get a lot of exercise throughout the day. Use training and playtime to accomplish this.

- **Set clear rules and boundaries for your dog**

    Do this from the very beginning. Set clear rules and boundaries, then introduce these to your dog one at a time through training.

If you see that your dog has forgotten any of these rules, correct them. If you have to, re-train your dog so that they will learn these things again. When your dog knows the rules and boundaries, you can enjoy a harmonious relationship with them.

- **Don't give your dog the same treatment as human beings**

Dogs aren't people, no matter how human-like they may seem. Therefore, you shouldn't treat your dog like a human nor should you expect them to respond like a human. This is why you should learn how to communicate in a way your dog will understand—by using clear, simple commands, hand signals, and body language.

- **Avoid scolding your dog**

No matter how many mistakes your dog makes throughout your training, keep in mind that they are doing their best. Never forget that dogs learn at different rates. If your dog takes a long time to learn a certain command, don't scold them for it. This type of punishment makes dog training a negative experience instead of a positive one.

Conclusion: Training Your Dog the Right Way

Fig. 21: Strong Bond. From Unsplash, by Humphrey Muleba, 2018, https://unsplash.com/photos/9MPZUdtw1RU/Copyright 2018 by Humphrey Muleba/Unsplash.

There you have it... everything you need to know about dog training. Whether you want to teach your dog the basic commands, potty-training, or other advanced tasks, you are now armed with the knowledge to do so. While the concepts involved in dog training are simple, the actual process is a long and tedious one. But as long as you remain patient, consistent, and committed to your dog's training, you will surely succeed.

From puppies to young dogs to adult dogs to senior dogs, you have learned how to approach training the right way. You have even learned which dog breeds are easy and difficult to train. This information is particularly useful if you are still trying to determine which dog breed to get. No matter what dog breed or stage of life your dog is in, it all boils down to communicating with your dog properly. Training is all about communication and finding out how your dog learns. While employing the different tips you have learned here, if you find that something doesn't work, change it! If you find that some of the techniques work effectively with your dog, then keep doing them.

Book 1 - The Complete Guide To Dog Training

As the trainer, you should be able to gauge how well your dog training is going. Don't expect your dog to tell you what's working and what isn't. Nor should you expect your dog to tell you what they like and what they don't. It's up to you as the owner to observe your canine and determine what can help improve and enrich your training sessions. Now that you have learned all the fundamental information to train your dog, it's time to start applying it. There's no time like the present to start your dog training journey. So, go ahead and begin strengthening your bond with your furry companion and teach them how to truly become man's best friend!

References

5 Cool Tricks to Teach Your Dog. (2019). Retrieved from https://www.purina.co.uk/dogs/key-life-stages/puppies/cool-dog-tricks-advanced-training

5 Dog Training Tips for a Great Family Dog. (2019). Retrieved from https://www.bupa.com.au/pet-insurance/dog-training-tips-family-dog/

5 Essential Commands You Can Teach Your Dog. (2018). Retrieved from https://www.cesarsway.com/5-essential-commands-you-can-teach-your-dog/

7 ways to discipline your dog. (2019). Retrieved from http://www.humansfordogs.com/2009/11/7-ways-to-discipline-your-dog.html

10 Advanced Security Training Tips for Personal Protection Dogs. (2019). Retrieved from https://www.expertsecuritytips.com/advanced-training-tips-for-dogs/

10 Best Training Tips. (2019). Retrieved from https://www.pedigree.com/dog-care/training/10-best-training-tips

10 Small Dog Training Tips. (2019). Retrieved from http://www.animalplanet.com/pets/top-10-tips-for-training-your-small-dog/

Adult and Senior Dog Training. (2019). Retrieved from https://www.purina.co.uk/dogs/behaviour-and-training/training-your-dog/ongoing-training-for-your-dog

Adult Dog Training – 10 Professional Trainer Tips. (2019). Retrieved from https://servicedogacademy.com/wp/free-dog-training-advice/adult-dog-training-10-professional-trainer-tips/

Advanced Dog Training. (2019). Retrieved from https://www.precision-dog-training.com/advanced-dog-training.html

Baker, L. How to Train Big Dogs. (2019). Retrieved from https://m.wikihow.com/Train-Big-Dogs

Basic Dog Obedience Training For Your Family Pet. (2019). Retrieved from https://www.precision-dog-training.com/basic-dog-obedience-training.html

Basic Dog Training Commands. (2019). Retrieved from https://www.purina.co.uk/dogs/behaviour-and-training/training-your-dog/basic-commands-for-your-dog

Basic Obedience Training for Dogs. (2019). Retrieved from https://www.instructables.com/id/Basic-Obedience-Training-for-Dogs/

Becker, M. 7 Tips for Training a Stubborn Dog. (2015). Retrieved from http://www.vetstreet.com/our-pet-experts/7-strategies-for-training-a-stubborn-dog

Becker, M. Dog Training 101: Essential Tools You'll Need. (2017). Retrieved from http://www.vetstreet.com/our-pet-experts/dog-training-101-essential-tools-youll-need

Becker, M. Your Dog Is Never Too Old for Training. (2012). Retrieved from http://www.vetstreet.com/our-pet-experts/your-dog-is-never-too-old-for-training

Bender, A. Teach an Old Dog New Tricks: 5 Training Tips. (2019). Retrieved from https://www.thesprucepets.com/training-tips-for-adult-dogs-1118253

Bender, A. Top 5 Ways to Use Positive Reinforcement to Reward a Dog. (2019). Retrieved from https://www.thesprucepets.com/ways-to-reward-a-dog-1118276

Bender, A. Training Small Dogs: What You Need to Know. (2019). Retrieved from https://www.thesprucepets.com/tips-for-training-small-dog-breeds-1118254

Bender, A. Why Positive Reinforcement Dog Training Works. (2019). Retrieved from https://www.thesprucepets.com/positive-reinforcement-dog-training-1118248

Berman, N. The 20 Most Difficult Dog Breeds to Train. (2017). Retrieved from https://puppytoob.com/20-difficult-dog-breeds-train/

Bolluyt, J. These Are the Dog Breeds That Are Notoriously Difficult to Train. (2018). Retrieved from https://www.cheatsheet.com/culture/dog-breeds-that-are-difficult-to-train.html/

Bolluyt, J. 18 of the Easiest Dog Breeds to Train. (2018). Retrieved from https://www.cheatsheet.com/culture/easiest-dog-breeds-to-train.html/

Bourne, S. How to Handle 6 Common Dog Behavior Problems. (2019). Retrieved from https://www.petcarerx.com/article/how-to-handle-6-common-dog-behavior-problems/1461

Can you teach an old dog new tricks: Here's how! (2019). Retrieved from https://tractive.com/blog/en/training-en/can-you-teach-an-old-dog-new-tricks-check-tips

Book 1 - The Complete Guide To Dog Training

Clark, M. How Much Training Is Too Much Training For Your Dog? (2019). Retrieved from https://dogtime.com/reference/dog-training/50681-much-training-much-training-dog

Clark, M. 7 Most Popular Dog Training Methods. (2019). Retrieved from https://dogtime.com/reference/dog-training/50743-7-popular-dog-training-methods

Cole, L. 6 Benefits of Obedience Training for Dogs. (2017). Retrieved from https://www.canidae.com/blog/2017/09/6-benefits-of-obedience-training-for-dogs/

Coren, S. Dogs Learn by Modeling the Behavior of Other Dogs. (2013). Retrieved from https://www.psychologytoday.com/us/blog/canine-corner/201301/dogs-learn-modeling-the-behavior-other-dogs

Day, L. 5 Steps to Create Your Puppy Training Schedule. (2019). Retrieved from https://pupbox.com/training/puppy-training-schedule/

Different Kinds Of Dog Training. (2015). Retrieved from https://www.cesarsway.com/different-kinds-of-dog-training/

Dog - Breed-specific behaviour. (2019). Retrieved from https://www.britannica.com/animal/dog/Breed-specific-behaviour

Dog Discipline – Should We Beat or Hit a Dog as Punishment? (2019). Retrieved from https://shibashake.com/dog/dog-discipline-punishment-beat-hit-dog

Dog Discipline: Does Hitting and Beating a Dog Work? (2019). Retrieved from https://pethelpful.com/dogs/An-Ear-for-an-Ear-Why-Biting-your-Dogs-Ear-Does-not-Work-aversive-techniques-forceful-punishment-do-not-work

Dog Training FAQ's. (2019). Retrieved from http://petkey.org/dog-training/trainingfaqs.aspx

Duno, S. Are You Making These 10 Training Mistakes? (2019). Retrieved from https://moderndogmagazine.com/articles/are-you-making-these-10-training-mistakes/29092

Easter, F. Do Different Dog Breeds Learn Differently? (2018). Retrieved from https://www.animalbehaviorcollege.com/blog/do-different-dog-breeds-learn-differently/

Effects of Dog Breed on Dog Training Effectiveness. (2014). Retrieved from https://unleashedjoy.com/breeds-harder-train-others

Elliott, P. How to Punish a Dog. (2019). Retrieved from https://www.wikihow.com/Punish-a-Dog

Elliott, P. How to Train an Adult Dog. (2019). Retrieved from https://www.wikihow.com/Train-an-Adult-Dog

Erb, H. Dog Training Tips: How to Train a Dog. (2017). Retrieved from https://www.akc.org/expert-advice/training/12-useful-dog-training-tips/

Fallis, C. Training a Senior Dog — And Other Valuable Advice. (2013). Retrieved from https://www.petful.com/behaviors/training-a-senior-dog/

Finlay, K. 12 Of The Easiest Dog Breeds To Train. (2019). Retrieved from https://iheartdogs.com/easiest-dog-breeds-to-train/

Freitag, S. 7 Reasons You Should Train Your Dog. (2016). Retrieved from https://www.theodysseyonline.com/7-reasons-you-should-train-your-dog

Gabbard, J. 10 Tips That Make Dog Training Easier. (2019). Retrieved

Book 1 - The Complete Guide To Dog Training

from https://www.puppyleaks.com/dog-training-easier/

Geier, E. The 7 Basic Must-Haves for Training Your Dog. (2019). Retrieved from https://www.rover.com/blog/basic-must-have-items-dog-training-in/

Get Started in Dog Training: Tips & Techniques for Beginners. (2019). Retrieved from https://www.thekennelclub.org.uk/training/get-started-in-dog-training/

Gibeault, S. Positive Rewards Dog Training Tips. (2018). Retrieved from https://www.akc.org/expert-advice/training/training-rewards/

Gibeault, S. The Importance of Training Your Senior Dog. (2017). Retrieved from https://www.akc.org/expert-advice/training/training-your-senior-dog/

Gigler, J. 5 Essential Dog Training Supplies. (2019). Retrieved from https://www.whole-dog-journal.com/training/5-essential-dog-training-supplies/

Harleman, J. How to Teach Your Old Dog New Tricks. (2019). Retrieved from https://www.rover.com/blog/teach-old-dog-new-tricks/

Heimbuch, J. 11 tricks you can teach a senior dog. (2015). Retrieved from https://www.mnn.com/family/pets/stories/11-tricks-you-can-teach-senior-dog

Helderman, J. 10 Bad Things That Happen When You Baby Your Dog. (2016). Retrieved from https://www.countryliving.com/life/kids-pets/g3433/bad-things-that-happen-when-you-spoil-your-dog/

Horwitz, D. Puppy Behavior and Training - Training Basics. (2019).

Retrieved from https://vcahospitals.com/know-your-pet/puppy-behavior-and-training-training-basics

House Training Adult Dogs. (2019). Retrieved from https://pets.webmd.com/dogs/guide/house-training-adult-dogs#1

Housetraining Adult Dogs: Training Tips And Techniques. (2019). Retrieved from https://dogtime.com/dog-health/general/360-housetraining-for-adults

House Training Your Puppy. (2019). Retrieved from https://pets.webmd.com/dogs/guide/house-training-your-puppy#1

How To Create A Puppy Schedule. (2015). Retrieved from https://www.cesarsway.com/how-to-create-a-puppy-schedule/

How to Potty Train a Puppy: Tips for New Pet Parents. (2019). Retrieved from https://www.petco.com/content/petco/PetcoStore/en_US/pet-services/resource-center/behavior-training/Tips-and-Tricks-for-Housetraining-a-Puppy-or-Dog.html

How To Train A Dog The Right Way. (2019). Retrieved from https://www.dog-training-excellence.com/how-to-train-a-dog.html

How to Train Your Dog & Top Training Tips. (2019). Retrieved from https://www.rspca.org.uk/adviceandwelfare/pets/dogs/training

Is it important to train my dog? What sort of training would you recommend? (2019). Retrieved from https://kb.rspca.org.au/knowledge-base/is-it-important-to-train-my-dog-what-sort-of-training-would-you-recommend/

Karetnick, J. Service Dogs 101: Everything You Need To Know About

Service Dogs. (2019). Retrieved from https://www.akc.org/expert-advice/training/service-dogs-101/

Kaschel, R. Are Some Breeds Harder to Train Than Others? (2019). Retrieved from https://www.petsafe.net/learn/are-some-breeds-harder-to-train-than-others

King, A. Training Tips: Can You Teach An Old Dog New Tricks? (2016). Retrieved from https://www.wideopenpets.com/can-you-teach-an-old-dog-new-tricks/

Kuras, A. 15 Helpful Dog Training Tips From The Experts. (2015). Retrieved from https://www.care.com/c/stories/6540/15-helpful-dog-training-tips-from-the-experts/

Lissa, S. 10 Common Dog Behavior Problems And How To Solve Them. (2019). Retrieved from https://www.dogviously.com/dog-behavior-problems/

Lotz, K. Top 10 Senior Dog Training Tips. (2019). Retrieved from https://iheartdogs.com/top-10-senior-dog-training-tips/

Lunchick, P. 5 Simple Commands You Should Teach Your Puppy. (2018). Retrieved from https://www.akc.org/expert-advice/training/teach-your-puppy-these-5-basic-commands/

Mattinson, P. Punishment In Dog Training. (2015). Retrieved from https://thehappypuppysite.com/punishment-in-dog-training/

McMahan, D. What your dog's bad behavior says about you. (2017). Retrieved from https://www.nbcnews.com/better/health/what-your-dog-s-bad-behavior-says-about-you-ncna795666

Meyers, H. Puppy Potty Training Schedule: A Timeline For Housebreaking Your Puppy. (2019). Retrieved from https://www.akc.org/expert-advice/training/potty-training-your-puppy-timeline-and-tips/

Millan, C. Starting Your Puppy Off Right! (2015). Retrieved from https://www.cesarsway.com/starting-your-puppy-off-right/

Miller, P. Training a Hyperactive Dog to Calm Down. (2019). Retrieved from https://www.whole-dog-journal.com/behavior/training-a-hyperactive-dog-to-calm-down/

Miller, P. Training An Older Dog - Whole Dog Journal. (2019). Retrieved from https://www.whole-dog-journal.com/training/training-an-older-dog/

Miller, P. Understanding Reward Based Dog Training. (2019). Retrieved from https://www.whole-dog-journal.com/training/understanding-reward-based-dog-training/

Miller, P. Young Dogs Learn From Older Well-Behaved Dogs. (2019). Retrieved from https://www.whole-dog-journal.com/training/young-dogs-learn-from-older-well-behaved-dogs/

Mouthing, Nipping and Play Biting in Adult Dogs. (2019). Retrieved from https://www.aspca.org/pet-care/dog-care/common-dog-behavior-issues/mouthing-nipping-and-play-biting-adult-dogs

New Tricks: Dog Training Tips For Puppy, Adult, & Senior Dogs. (2019). Retrieved from https://www.justrightpetfood.com/blog/dog-training-tips-for-puppy-adult-senior-dogs

Obedience Training for Dogs. (2019). Retrieved from https://pets.webmd.com/dogs/guide/dog-training-obedience-training-for-dogs#1

Ohlms, S. 7 tips for training your dog, from a Marine who trained dogs to sniff out bombs. (2019). Retrieved from https://www.businessinsider.com/7-tips-for-training-your-dog-from-a-military-dog-handler-2019-5#7-not-every-dog-is-going-to-be-able-to-learn-every-task-7

Parks, S. The First 5 Things to Teach Your New Puppy (and When to Start). (2019). Retrieved from https://www.rover.com/blog/can-train-puppy-first-day-heres/

Pennings, S. Precision K9 Dog Training - Obedience for the Family Dog. (2019). Retrieved from https://www.precision-dog-training.com/

Phenix, A. 5 Training Tips for Your Working Dog Breed. (2017). Retrieved from https://www.dogster.com/dog-training/training-tips-for-your-working-dog-breed

Phenix, A. How to Prep for the First Two Months of Puppy Training. (2019). Retrieved from https://www.dogster.com/puppies/first-two-months-puppy-training-prep

Re-Housetraining Your Adult Dog. (2019). Retrieved from https://www.paws.org/library/dogs/training/re-housetraining/

Rollins, J. How to Discipline a Puppy or Dog: Effectively Punishing Your Dog. (2016). Retrieved from https://www.petexpertise.com/dog-training-article-using-physical-punishments/

Secrets To Housebreaking Adult Dogs. (2019). Retrieved from https://www.cesarsway.com/secrets-to-housebreaking-adult-dogs/

Shea, T. How to Train a Puppy: The First 8 Things You Need to Do. (2019). Retrieved from https://www.rd.com/advice/pets/how-to-train-your-puppy/

Stregowski, J. How to Solve 10 of the Biggest Dog Behavior Problems. (2019). Retrieved from https://www.thesprucepets.com/common-dog-behavior-problems-1118278

Stregowski, J. This Step by Step Guide Can Help You Completely Train Your Dog. (2019). Retrieved from https://www.thesprucepets.com/steps-to-train-your-dog-1118273

Teaching old dogs new tricks. (2019). Retrieved from https://dogtime.com/lifestyle/dog-activities/1161-training-adult-senior-dogs-aaha

The Basics of Puppy Training. (2019). Retrieved from https://www.zooplus.co.uk/magazine/dog/dog-training/basics-puppy-training

The Do's And Don'ts Of Positive Reinforcement. (2019). Retrieved from https://www.cesarsway.com/the-dos-and-donts-of-positive-reinforcement/

The Importance of Training Your Dog. (2019). Retrieved from https://www.greenacreskennel.com/dog-behavior-and-training/the-importance-of-training-your-dog.html

Theis, T., & Conway, K. Top Ten Dog Training Tips. (2019). Retrieved from https://www.petfinder.com/dogs/dog-training/dog-training-tips/

Tips for Solving Common Behavior Problems. (2019). Retrieved from https://www.nylabone.com/dog101/tips-for-solving-common-behavior-problems

Top 5 Best Dog Training Super Tips For Beginners. (2019). Retrieved from https://sitstay.com/blogs/good-dog-blog/95094215-top-5-best-dog-training-super-tips-for-beginners

Book 1 - The Complete Guide To Dog Training

Top 10 Hardest Dog Breeds to Train. (2017). Retrieved from https://canna-pet.com/top-10-hardest-dog-breeds-train/

Training An Older Dog - You Can Teach An Old Dog New Tricks!. (2019). Retrieved from https://www.fidosavvy.com/training-an-older-dog.html

Training Older Dogs - Tips & Advice That Work. (2019). Retrieved from https://www.seniortailwaggers.com/training-older-dogs/

Treat training for dogs. (2019). Retrieved from https://www.cesarsway.com/tricks-for-treats-training-your-dog-with-food/

Vuckovic, A. How to Discipline a Dog Without Hitting, Advice and Tips. (2017). Retrieved from https://petcube.com/blog/dog-training/

Waggener, N. Five Basic Obedience Commands Your Dog Should Learn. (2018). Retrieved from https://www.southbostonanimalhospital.com/blog/five-basic-obedience-commands-your-dog-should-learn

Welton, M. Puppy Training Schedule: What to Teach Puppies, and When. (2019). Retrieved from https://www.yourpurebredpuppy.com/training/articles/puppy-training-schedule.html

White, P. MASTERCLASSES; DOG TRAINING; How to teach young dogs new tricks. (1995). Retrieved from https://www.independent.co.uk/arts-entertainment/masterclasses-dog-training-how-to-teach-young-dogs-new-tricks-1526258.html

Wondra, S. Large Dog Training Tips. (2019). Retrieved from https://www.petcarerx.com/article/large-dog-training-tips/862

Woods, J. Puppy Training Tips: 45 Dog Experts Share Their Secrets. (2018). Retrieved from https://www.allthingsdogs.com/puppy-training-tips/

Wright, M. Philosophy. (2019). Retrieved from https://www.argostraining.com/positive-dog-training-philosophy/

BOOK 2

E-COLLAR TRAINING STEP-BY-STEP

A How-To Innovative Guide to Positively Train Your Dog through E- collars; Tips and Tricks and Effective Techniques for Different Species of Dogs

PAUL DAVIS

Introduction

You are at your local animal shelter looking through the cages. You walk by slowly, talking to each dog. "Hey there, how are you doing today?" you say to each one of the dogs as you read the little notecard of information attached to the cage. It's so hard to choose one dog to bring home. You want to bring them all home. But you know that you need to pick the best dog for yourself, your family, and think of the dog.

Living in an apartment, you realize a smaller dog is best. It will get exercise by running around the apartment easier than a larger dog will. You also know you need a dog that is comfortable around kids, won't get lonely easily, and is easily trainable. With you and your husband working full-time jobs, you can't be home all the time for your new pup.

Through your research, you know many breeds of smaller dogs are easily trainable. Walking by a cage, you see a Miniature Schnauzer named Phillip. You smile at Phillip as you read information about him. He is three years old, neutered, enjoys children, and has received basic training. Through your conversation with the shelter employee, you find out that the employees are training Phillip with an e-collar or electronic collar. You know very little about this device, so the employee shows you how it works.

This book is going to take you through everything you need to know about e-collar training and your dog. You will not only learn the basics of e-collar training, but you'll learn how to help your dog thrive. Diving into Chapter 1, you will get basic information, such as easily trainable dog breeds, what factors to think of when you are choosing the best dog for you, how to care for your dog, and where to go to choose your dog.

Chapter 2 will give you information on e-collar basics. After learning what an e-collar is, you can discover the types of e-collars and accessories available for your dog. This chapter will also give you a step-by-step guide on how to use an e-collar. Of course, you can't learn how to use an e-collar without learning about the safety of e-collars. If you are still unsure if e-collar is the way to go, Chapter 2 debunks several myths that people have about e-collars.

Deciding that you want the e-collar, you wonder how it will affect your dog. Is there a certain e-collar you should choose? What will your dog's reaction be? Instead of wondering about these questions, it's time to read Chapter 3. In this chapter, you will learn about the fundamental five when it comes to picking the right e-collar for your dog. You will also read about a couple of case examples when it comes to the unintended consequences of an e-collar your dog can have if you do not train them properly.

Chapter 4 gives you everything you need to know before you begin training. You will receive general tips on training a dog at any age, information on obedience classes, and how your dog's age matters when it comes to training. To give you the greatest amount of information possible so you are prepared to start training your dog, this chapter gives you tips for training your puppy and tips for training an older dog.

Finally, in Chapter 5, you can let the training begin. This chapter starts with giving you tips to prepare your dog for training. You will then get a step by step guide on how to train your dog with the basic commands, such as bed training, home base training, sitting, laying down, and getting down.

Chapter 6 goes a little further into training by looking at the training levels you will reach if you enroll your dog in obedience school. Of course, you can always focus on training your dog through these levels in your home. One of the best parts of this chapter is that it gives you some real life training examples so you know what to do when you are practicing training your dog in a real setting, such as bringing them on a car ride or to a dog-friendly store. You will also receive information on how to handle aggressiveness from dogs and some intermediate level tricks, such as teaching your dog to play fetch.

The advanced training shows up in Chapter 7. Not only is the popular question of "Will I ever stop training my dog" be answered, but you will read tips about agility training and hunting. You can also learn a couple of fun tricks to see if your dog would be great at agility training.

What are the most common mistakes dog owners make and how you can pay attention so you don't make the same mistakes? You can find out all this information in Chapter 8. You will learn how discipline doesn't really help your dog, though most people incorporate time out when their dog does something like scratch up the wall.

This book will round off with Chapter 9, which discusses a number of common questions people have about dog training. Of course, you will get the answers to these questions!

Now that I have given you the rundown for what you are about to learn, it's time to dive into the wonderful information!

Chapter 1: Choosing the Right Dog for You

Choosing the right dog can be a big decision. While sometimes it seems that the right dog just "falls into your lap," other times you search for days or weeks. There are a lot of factors that influence your decision in getting a dog. For example, if you don't have a lot of money to put toward a dog, you look for a free dog or one from the shelter that is already fixed. If you live in an apartment, you may feel it is easier to get a small dog.

Tips for Choosing the Best Dog for You

With over 200 breeds of dogs, it's impossible to discuss all of them. Instead, I want to focus on tips that will help you choose the best dog for you.

Dogs Are Not Cheap

One factor to consider is how much money you can put toward a dog. While they are cheaper than children, they still need to go to the vet, they need food, toys, a crate, a leash, bowls for their food and water, grooming, and training tools or classes. Just because you got a dog for free from a social media site or your neighbor, does not mean that you will never put money toward your dog. Plus, most people like to spoil their dog with a sweater, especially if they live in colder temperatures.

Plan for Your Dog's Arrival

It is important to not just bring a dog home without any preparation. You want to prepare a place for your dog just like you would any member of your family. Just like humans, animals have feelings. They also need to feel a certain way to grow within their new home.

If you have an older dog, you won't have to worry about "puppy proofing" your home. However, you will still want to ensure that you pick up anything the dog can harm themselves with. Even senior dogs will find interest in a screwdriver or want to taste the chocolate within their reach.

If you are bringing home a puppy, you will want to make sure that you puppy proof your entire home—you will be amazed at what a puppy can get into. For instance, you will want to ensure all your valuables are not in the puppy's reach. Remember, your puppy will be full of energy and into jumping. You will also want to secure any wires because puppies love to chew on anything.

You also need to think about what the dog needs. Some of the factors to think about are:

- *Security.* Just like people, one of the biggest ways to make a dog feel at home is to make them feel safe.
- *Love.* You will need to spend time with your dog and make sure they are adjusting well. They will need a lot of physical touch and attention. This will help calm their fears and know they are a welcomed member of the family.
- *Calming environment.* Dogs, especially puppies, become overwhelmed when there is a lot of chaos. They don't know what to do or how to handle it. This can make them act out, causing other problems.
- *Their space.* Just like people, dogs need their space. They need an area they can call their own. This is an area where they should feel safe and can have some dog time, just like humans need "me time."
- *They need supervision.* If you have a child and get a puppy, you will find a lot of common characteristics between the two, especially if you have a toddler and a puppy. While older dogs don't need

as much supervision, a puppy is going to need a lot of attention and supervision. Puppies can become easily wound up and don't understand they can hurt someone if they jump on them. Therefore, watch your pup when they are playing to ensure everyone stays happy and safe.

You Need to Exercise Your Dog

If you are not one for walking, jogging or running, you will need to think about a dog walker or start exercising. Dogs need a lot of exercise, including walking every day. Dogs will need room so they can run, play, and jump. Therefore, you should always think twice about having a large dog in a small apartment where they rarely get to go outside.

Fortunately, if you are not one for exercise but still want a dog, there are breeds of dogs that require little exercise. For example, senior dogs will still need to go outside for a walk, but they are tamer and have lost a lot of energy. Smaller dogs are often okay to have if you are not huge on outdoor exercise because your home will give them room to play around and exercise.

Most Dogs Live for 10 Years or More

When you choose a dog, you want to keep in mind that you will most likely have this dog for over a decade. While this is usually a plus when it comes to picking a dog, some people are not interested in having a dog for years for various reasons. If this is the case, one of the best steps to take is to adopt an older dog. While you will need to factor in health problems, they might be the best choice if you only want a dog for a few years.

At the same time, you might want a dog who is known for a long lifespan. For example, you are getting a dog for your 10-year-old child to teach them responsibility. It takes years for most children to learn how to be responsible. At the same time, you will want a younger dog because you want your child and the dog to spend years together.

Most Dogs Require Grooming

The amount of grooming your dog needs will vary depending on the breed. Some dogs tend to need their fur trimmed often, while the fur on other breeds is fine without trimming. If you have a short-haired dog, you won't need to send them to the groomers as often. However, long-haired dogs require regular grooming and everyday brushing.

All dogs will need to get a bath from time to time. You need to wash their coat with special soap that is not only safe for the dogs, but also valuable for their fur. Along with fur comes shedding. Most dog breeds shed, but short-haired dogs are going to shed more than long-haired dogs.

A Puppy Doesn't Stay Small Forever

Puppies are adorable. Most people seem to want a puppy for several reasons. They are small, will lay in your lap, playful, and silly. Unfortunately, puppies do not stay small forever. Before you decide on a puppy because of its size, do a little research on the breed to see how big they can grow. Even puppies who seem on the smaller side for their breed can grow into large dogs. Don't let the adorable puppy size fool you!

Can You Meet the Parents?

Some people, especially people who want a show dog, will only take home a puppy if they can meet the puppy's parents. While this might not be important to you, it is something to consider if you have children, want to train your dog to become a show dog, or simply want to meet the parents. Parents can tell you a lot about how your puppy could grow up. But you always need to remember, a lot of your dog's temperament is going to deal with how you treat them.

Choosing the Best Breed for Training

It's true, some dogs are easily trainable. If you know that you will train your dog and don't want too many struggles with it, you will want to look at breeds that are easy to train.

What Makes a Dog Easily Trainable?

There are many factors that you need to think of when looking for an easily trainable dog.

- *Are they easily distracted?* Some dogs are more distractible than other dogs. For example, if you want a dog for hunting, you will not want to get a dog that is going to get distracted by leaves rustling in the trees or blowing in the wind. This can cause you to lose your target. You want a dog that is going to be aware of their surroundings yet stays focused on its target. Strong concentration also helps when you are training. Your dog will stay focused on what you are telling them and your hand signals.
- *What is the dog's personality?* Each breed has a distinct personality. Of course, each dog is a bit different than the others, but each breed has certain personality characteristics. For example, some breeds are going to cooperate better than other breeds.
- *What is the dog's instinct drives?* Each breed has certain instincts that drive them. This means these drives take over the dog's other instincts. For example, bloodhounds are ruled by their nose. Therefore, what they hear and see at the same time they are smelling something is not going to matter as much.

Best Breeds for Training

Miniature Schnauzer

As members of the Terrier group, Miniature Schnauzer's are not considered easily trainable dogs. This is because most Terrier dogs are harder to train than other breeds. However, Miniature Schnauzer's enjoy pleasing their owners and will work hard to accomplish a task. They are fearless, playful, and have a better trainability rate than any other dog in the Terrier family.

These small dogs can go anywhere you are. They can roam around on a farm in the country or remain content in an apartment. They have a spunky personality and make people laugh with their human-like expressions. Miniature Schnauzer gets along well with other dogs and children. They tend to make great watchdogs.

Doberman Pinscher

The Doberman Pinscher is a loyal and fearless companion. They are one of the breeds that are used for police and military training. They listen well, have great attention to detail, and the ability to concentrate on a task. Doberman Pinschers understand their commands, are quick to learn, and retain a lot of information.

Doberman Pinschers are not commonly chosen to be companions because they have a stereotype of being ruthless and known to attack. In reality, these dogs are cuddly, love to play, and friendly. They will only attack when they are threatened, like most dogs, or are trained and told to attack.

Doberman owners are honest about the stereotype their companions deal with on a daily basis. Walking your Doberman down the street, you start to notice people crossing the street ahead of you. They are watching how close you get as they cross. A few minutes later, you sit on a park bench to get a frisbee out of your bag. The guy next to you says, "I heard those dogs would attack their owner after their brain stops growing. Aren't you afraid of that dog attacking your neck?" You look at the person, get down on your knees, and say, "Come here, come give me a hug." You Doberman companion gently places both of their front paws on your shoulder, and you embrace them in a hug. Looking back at the person you say, "Not at all. They are big teddy bears."

Andrew Wildesen wrote an article dedicated to the stereotype Doberman Pinschers and their owners face. Wildesen stated that Dobermans are similar to every other dog–they can be trained to attack. However, they won't innately attack. Instead, they will bring more love and happiness into your life. Wildesen states, "The true Doberman is a lover with loads of enthusiasm... They are goofy and are guaranteed to make you laugh... they are tremendously loyal and love to cuddle... yes, Dobermans... are serious CUDDLE BUGS!" (Wildesen, n.d.).

One of the factors to be aware of with a Doberman is you need to put a lot of effort into their training. If you are looking for a dog that is easily trainable but doesn't require a lot of time, you will want to look at a different breed. You need to provide constant leadership and training for the Doberman breed to thrive.

Poodle

Poodles are common dogs chosen for many reasons. First, they are smaller in size, allowing them to adapt to various types of living situations. Poodles come in three different sizes: toy, miniature, and standard. Second, they are typically energetic and great with kids. Third, they are on every list that describes easily trainable dogs.

People love to train Poodles because they are extremely smart. However, some owners say they are too smart for their own good as they are known to outwit their trainers! They learn their tricks quickly and are eager to please their owner. It doesn't matter what size of Poodle you choose, they will all do their best to surprise you with the tricks you teach them.

Rottweiler

The Rottweiler breed is similar to Doberman Pinschers. They are known for their aggressive nature, but it is all in their training. On top of this, Rottweilers are easily trainable dogs because it is in their nature. In fact, they crave training and it is essential for their happiness.

Rottweilers are great to use as service and police dogs. However, they not only need basic training, they also need social training. This type of training goes beyond sitting and shaking hands as they like to be challenged.

German Shepherd

The German Shepherd is the Jack-of-all-trades dog. They are easy to train because they understand commands. They remember the tricks they are taught and need jobs to keep themselves happy. Furthermore, they are willing to learn and want to please you. They will go above and beyond the call of duty. German Shepherds tend to work tirelessly and are great for military and police work.

Like most dogs, German Shepherds need a lot of exercises. You will need to keep their mind going as well as their body. You don't need to wait too long to start training these dogs. They can understand the basic commands as young as eight weeks old.

Havanese

Another dog that does great with apartment living is the Havanese. They are highly intelligent, making them easily trainable. Furthermore, these dogs are full of energy and love the outdoors. This means that you will need to make sure your Havanese can get outside to let run off some of their energy. At the same time, because of their smaller size, they will get a lot of running done indoors.

Like most other dogs, the Havanese's coat needs to be brushed a couple of times a week. They don't tend to shed, so are great hypoallergenic dogs. When training a Havanese, positive reinforcement is your best procedure. They are sensitive dogs and quickly understand your tone of voice. Once they realize the tones, they will judge how you feel about them and their actions by your tone.

Shetland Sheepdog

The Shetland Sheepdog is a competitive dog. They enjoy dog sports, are playful, and affectionate. They love learning new behaviors as it keeps their mind and body active. If you have a Shetland Sheepdog, you should enroll them in obedience or agility classes because they will thrive with this additional training. Like the German Shepherds, Shetland Sheepdogs want to please people. This causes them to work hard to learn new tasks. Because of their high intelligence, they learn and retain new information easily.

Shetland Sheepdogs are great to have in the country but can adapt to apartment living. However, they are extremely active and can grow about 13 to 16 inches at the shoulder when standing. They are sensitive dogs and don't tend to bark unless they feel danger or a stranger is approaching. Therefore, they make great watchdogs.

The Papillon Club

The Papillon is an active little dog who needs to be kept busy. They don't get much taller than 11 inches at the shoulder. If you choose this breed, you will want to think about an active training schedule. The training you give the Papillon should mentally motivate them as well as physically. They are curious dogs and quick to learn.

Unlike some breeds, the Papillon learns from each experience, and they remember everything. Therefore, you need to maintain consistency as a trainer. If you are not consistent, your Papillon is not going to thrive. They will become confused and seem like they lack skills, including potty training.

Border Terrier

Border Terriers are a smaller dog that can easily adapt to any living situation. While they are not the most popular dogs, they are great dogs for families. They love children, love to learn, and great dogs for training. Because border terriers enjoy agility classes, they are perfect dogs for easy training.

Australian Shepherd

The Australian Shepherd is a medium-sized dog, which makes apartment living a bit of a challenge for them. They don't like to spend a lot of time indoors. Australian Shepherds thrive best when they have a large area to run around, play, and learn.

Australian Shepherds are known for their herding instincts. For example, many people will train them for herding sheep or cows from one pasture to the other. However, some owners have also found their dogs trying to herd a group of children. The dogs will not hurt the children, but younger children can become confused and frightened when an Australian Shepherd is trying to herd them into a designated area.

The best home for an Australian shepherd is with someone who can devote a lot of time in training them. They are dogs that need a lot of love and a firm voice to keep them in line. They can lose sight of what they are supposed to do because of their boundless energy.

While some breeds are fine with training starting later in life, Australian Shepherds need to start as puppies. They are happy when their owners take time to train them and can work off much of their energy.

Golden Retriever

The Golden Retriever is one of the most popular breeds. They are known for their beauty and hunting skills. They are also used as seeing-eye dogs and for search and rescue missions. They are willing to learn and enjoy competitive events. A Golden Retriever is a dog that you would enroll in an obedience class because they enjoy this type of setting.

Golden Retrievers are family dogs that are eager to please. They are highly intelligent, making them easily trainable. They have a happy and playful approach to life, making them energetic and powerful dogs for outdoor activities. Because they are a medium-sized dog, they don't enjoy apartment life well. However, if you make sure they exercise daily and take them outside, they will do fine in an apartment.

Where to Go?

Now that you are briefed on some of the most trainable dog breeds, you might be asking yourself where to go. Thanks to the internet and social media, the places you can find a dog of your choosing is right at your fingertips.

Adoption

One of the first places to go is a local animal shelter or rescue group. There are a variety of dogs you can find in a shelter. Some places are mainly for purebreds, but most take in any dogs that need a place to stay until they are adopted.

One of the biggest challenges for people when adopting a dog is the unknown history and how to make the dog comfortable in their new home. To help your dog adjust to their new life, here are a few tips to follow.

- *Don't expect too much right away.* The new member of your family needs time to adjust. You shouldn't start to train them right away. Your main focus for the first few days to a couple of weeks is making them comfortable.

- *Understand they have emotional trauma.* Most dogs that come from a shelter have some type of emotional trauma, and you need to help them heal. This is going to take time. The dog was abandoned or got lost. They don't understand what happened to their companion, who they still love and miss. Plus, their time in the shelter can cause emotional damage, even if they were treated well. Be patient as you work through this trauma with them; it will take years to repair. Unfortunately, some trauma will remain with the dog.
- *The dog might act differently in the shelter.* If you come across a dog you fall in love with, but they act scared and bark a lot doesn't mean they will act this way in your home. The shelter makes them anxious, causing them to react in this way. Don't believe they will always act in that way. Once you bring them home and they become comfortable, the dog will relax and act typical for their breed.
- *Start a routine right away.* One way to help the dog feel comfortable in their new environment is to establish a routine immediately. This doesn't mean you start training them right away. Instead, you will feed them at a regular time, show them where their water is, place their bed in a certain area, go to bed at a specific time, get up at a regular time, and play with them. If you struggle with routines yourself, get a planner and what down what to do when with your new dog. Be as consistent as possible because they will start to trust what is happening next. If their routine changes, it can make them anxious.

Pet Store

I won't deny this—there is a lot of debate about pet stores. Some pet stores do receive dogs from puppy mills, but not all. Do thorough research on your pet store before you decide to get a dog there. Go to visit the possible animal a couple of times and get to know the employees. While they can't hold the dog you are interested in for you, knowing the background of the store will help you understand the dog better.

Some pet store will advertise that they are "puppy friendly," meaning they don't use puppy mills as a resource. While this can be true, remember anything can be said through advertising. Through research, you should find out where the puppies come from as some stores do take homeless dogs.

Responsible Breeders

Some people breed dogs. These dogs are usually purebred and well taken care of. Of course, you will still want to get to know the breeders. If you can, talk to other people who have gotten puppies from them. Interview the breeders and get to know the breeders. Ask them why they decided to breed and how long they've had the parents.

Social Media

There are a lot of social media sites that will sell puppies. Usually, the people have a dog that became pregnant and they can't care for all the puppies. Sometimes they are giving away the dogs and sometimes asking for money. Typically, they don't ask for a lot of money, but it depends on the breed. If you feel that they are asking too much for the dog, do a little research. They might be part of a puppy mill.

You always want to be cautious of online social media sales. While most occur locally, there are people willing to ship the dog if you send the money to them. This can easily be a scam, just like any smaller business stating the same thing. Always do your research if you don't know the people.

Tips to Keep in Mind Before Bringing Your Dog Home

It is always a bit challenging for some of us to say we are searching for our "best" dog. If you love dogs and want to take one home, you are going to be fine with getting a free puppy out of a cardboard box from your neighbor's home. Of course, there is nothing wrong with this. But, for many people, finding the best dog is necessary for several reasons. Whether you are looking at your neighbor's new litter of puppies or thinking of heading to the animal shelter, here are a few tips to help you along the way.

Be Cautious of Internet Sites

You are searching online for your Australian Shepherd. You know there isn't an Australian Shepherd available locally. However, you notice this small business called "Worldwide Love For Dogs Rescue." As you read about the newly established company, you come to learn that they will talk to you through video chat where you can see your potential dog. You can also receive email information about your dog's personality before making the final purchase. Then, once you send them the money, they will transport your dog to you. Sounds like a pretty good deal if you find the perfect Australian Shepherd for your home, right? This can be a good deal, but it can also be a scam.

One factor to consider is what do you really know about the company? You should always read up on any reviews, whether they are positive or negative. Do a little research on the company. Are they establish as an LLC or another type of organization? You can often find this information online. Do they keep their business clean?

Another factor to consider is how well they treat their dogs. People can easily tell you how they treat animals in a way that makes you believe they take care of their dogs. But you never truly know until you see the dog or do a little research on the company.

It is always a good idea to do a background check on the person you are thinking of doing business with. While you might have to pay for a background check, there is information you can find for free through research. Remember, people can say anything in advertising. No one is going to tell you that they have 125 dogs caged in unsanitary conditions because they are running a puppy mill or into dog fighting.

Visit the Dog on Site

Another reason many people will caution anyone against getting a dog through an online store where they are shipped is because visiting the dog on site is 100% helpful. There are hardly any downsides to visiting the dog before you decide to take them home. The only real downside is you are going to fall in love with the dog and have to wait until any paperwork and your background check clears or you will fall in love with all the dogs and can't bring them all home!

Seeing where the dog sleeps and temporarily lives will allow you to understand the dog. You will see firsthand how well they eat and interact with other people and animals. You will also make the dog more comfortable when you bring them home. Dogs can be nervous when they are chosen as they don't understand what is happening. While they might seem excited, like any animal, they will be cautious of their new surroundings and home. Meeting the dog in their home can help them feel more comfortable during transportation and their new living situation.

Poor Breeding Practices and Living Conditions

Visiting the site will help you see how well the dog's temporary residence is taken care of. If you walk into a building and see dozens or hundreds of dogs, some that are kept in the same cage, the building smells bad, and the dogs seems scared, ungroomed, thin, and not cared for (i.e., water might be dirty or water bowl empty) you might be in the middle of a puppy mill situation. It's important to note that animal shelters have standards they need to ensure they follow. Puppy mills are typically illegal and aren't kept up to standards.

If you ever find yourself in this situation, the best step to take is to try to get some background information. Are they running an animal shelter that is registered as a business? Closely check the conditions of the dogs. If you notice they don't give you a lot of information or don't answer your questions, contact the police department or a local animal welfare agency. It's always better to ensure that the dogs are well taken care of than to allow a puppy mill to continue.

While you might want to get one of the puppies to bring home for a better life, you need to be aware that dogs from puppy mills are often sick. They can carry diseases from the unsanitary conditions and need to be taken to a veterinarian immediately.

Chapter 2: E-Collar Basics

When you look at the instructions for an e-collar, you may feel confident you have all the information you need. Unfortunately, you are wrong. While companies give you detailed instructions when you purchase their e-collar, it can still be confusing for many reasons. First, each breed is going to react differently to the e-collar. This will happen because of their personality and how trainable they are. For example, some dogs will struggle with the e-collar because they are more independent and don't care to please their owners like other breeds. Second, each button is going to make the dog feel a different way, making your dog react in different ways to the e-collar. This can throw you off as you fear you might be doing something wrong.

Don't allow the way your dog reacts to the buttons throw you off. This is going to throw your dog off as well. Dogs can sense how you feel. For example, if you push a button on the e-collar and you notice them take a step back, you start to feel worried you hurt them. Your dog will sense this, causing them more uneasiness with the e-collar and training in general.

What Is the E-Collar?

A little over 30 years ago, the first electric dog collar made an appearance. Initially used to train hunting dogs, people became excited and worried about how the collar affected their dog. Some people felt they are harmful while other people believed they are a great training tool. The first e-collar only delivered one stimulation level. Today, there are three stimulation levels–high, medium, and low–and various levels of intensity within these stimulation levels for many e-collars. It is safe to say that over the last three decades, e-collars are safer, made for all types of dogs, and a perfect training tool.

The e-collar is a collar that has a radio receiver attached to it. The radio will receive a signal from the transmitter. This signal will give your dog a small electrical shock, similar to how you feel static electricity. This startles your dog enough to know that what they are doing is not appropriate. Over time, you will train your dog through the e-collar to know what behavior is right and wrong.

Most people choose e-collars because they don't want to train their dog with a leash. Another reason people choose e-collars is that they can train their dog from a small distance. For example, you are at the park with your dog when they start running off. Instead of yelling at your dog, you press a button. This alerts your dog to stop running and come back to you. Of course, at this point you have trained your dog to respond to the shock by looking at you to see what you want. For instance, you will motion them to come back. If you want them to sit, you will motion them to sit.

You can purchase a e-collar in most stores or online. If you've never worked with an e-collar, you might want to go to your local pet store. The employees there will help you understand how the e-collar works and answer any questions you have.

Types of E-Collars

When searching for the best e-collar for your dog, you need to think of why you're training. E-collar aren't always used for basic training, but some are focused for yard use. There are also e-collars for hunting, no barking, and working dogs.

Yard Training

Yard training e-collars is mainly for your pets, but some can be used for other training purposes. They are some of the most common e-collars and give you a lot of options.

- *EZ 900 Easy Educator* is an e-collar that focuses on safety for your dog. Its boosting level reaches to 60 and stimulation level to 100. It's considered the most humane e-collar on the market. This e-collar is a smaller size, making it great for any size dog. Many people who purchased this product say you won't see the typical head jerking coming from your dog. You can use the EZ 900 up to ½ a mile and can be used for hunting and working training as well. Another benefit of this collar is it's one of the cheaper ones on the market.
- *ET-302 Zen Educator* is another humane collar that can reach about a ½ mile. The stimulation level reaches to 100 and holds boosting levels from 1 to 60. It includes a collar to help you find your dog in the dark and is waterproof. This collar is sized for any dog and has two receivers, but is specifically for yard use.
- *ET-402 Educator* has stimulation levels from 1 to 100 and boosting levels from 1 to 60. This e-collar can reach up to ¾ of a mile and is great for emergency situations. It's waterproof and includes a light. Any size dog can wear this collar, but it's specific to yard training.
- The *FT-330 Finger Trainer Educator* is one of the newer models that can reach up to ½ a mile. It's a smaller collar, but holds enough stimulation for any sized dog. This collar holds most of the same specifics of the other collars, but includes a remote finger button.
- *ME-300 Micro Educator* works best for smaller dogs. It's a smaller version of the RX-090 Mini Educator Receiver but holds the same features. Its range is a ⅓ mile and charges quicker than other e-collars.

Book 2 - E-Collar Training Step-By-Step

Hunting Training

While a few yard training e-collars are used for hunting as well, many professionals feel it's better to get a specific hunting collar.

- *UL-1200 Upland Dog Trainer* is one of the cheaper hunting e-collars, but it's also older than some models. Its stimulation and boosting levels are the same as other e-collars. You won't want to use this e-collar on smaller dogs because of its size. One benefit of this e-collar is it can be used for yard and working training as well.
- *WF-1202 Waterfowler* is a perfect hunting training e-collar. It's the best on the market and can reach up to one mile. While this e-collar is useful for any type of training, it's only for medium and large dogs.

Work Training

Work training e-collars are perfect for dogs with certain jobs and performances. They are sometimes used in seeing-eye dog training or dogs preparing for dog shows.

- *ET-802 Dog Remote Trainer* is a newer version of the ET-800 Dog Remote Trainer. It holds the same type of stimulation and boosting levels as other e-collars. This e-collar is usable for any type of training, but only used on medium or large dogs. It's weight and size is too much for a smaller dog to handle. It can reach up to one mile.
- The *K9-402 K9 Handler* can reach up to ¾ of a mile. It's one of the smallest working dog training e-collars, making it perfect for any sized dog. You can also use this e-collar for yard or hunting training.
- *Pro Educator 900* is an older e-collar built with safety in mind. It's smaller in size, so perfect for any type of dog. It reaches up to ½ a mile and is also great to use as a no-bark collar.

No-Bark E-Collar

There aren't a lot of e-collars specific for no barking as you can typically use any type of e-collar for this training. However, the main no bark e-collar is the *BP-504 Anti Bark Collar*. It's newer on the market, has nine stimulation levels and three sensitivity levels. This e-collar is great for no barking because it comes with a warning sound your dog will quickly understand. This e-collar should not be used for any dog under five pounds.

E-Collar Accessories

Each e-collar will come with a manual and it's necessary pieces, such as the transmitter, battery charger, lanyard, contact point tool, e-collar, strap, and extra contact points. You can also purchase various accessories to go with your e-collar. For instance, many people like to purchase another contact point tool as it allows for more than one person to control the dog. As long as you are both on the same page, this won't confuse your dog. You just want to ensure that you are both aware of what type of training you're focusing on. Other accessories include:

- A carrying case for all your dog training tools. Some people don't keep the e-collar on their dog as it's only for training purposes. For example, if you are training your dog to hunt, you don't need to have the e-collar on your dog when you're at home.
- A dummy e-collar. Some people purchase this accessory because it makes your dog believe the collar is still on when it's charging. If you are not constantly training your dog, this isn't necessary unless you want to test your dog on their training.
- The Educator Gear Keeper is great for anyone who is training their dog on the go. You can attach this device to your belt loop or purse to give you quick access. It allows you to put away the transmitter instead of keeping it in your hand constantly.

- Educator bungee straps are popular because they come in a variety of colors. They come in 33-inch and 37-inch lengths but can be cut down to size. They are also extremely comfortable for your dog and easy to clean. Unfortunately, they are not the best for smaller dogs.

How to Use an E-Collar

E-collars are not meant to be used as a source of punishment. They are meant to deter the dog from unsafe and negative behavior. The purpose is an e-collar will train the dog that their negative behavior gives them an uncomfortable shock. The dog will remember this sock and stop the behavior. This theory will only work if you don't allow your dog to know you control the collar and you allow them to become comfortable with the collar.

Parts and Terminology

- **Stimulation level.** The best e-collars to get are the ones that have at least 100 levels of stimulation. The more levels you have, the easier it is to train and select the best shock for your dog.
- **Power Output.** Do not assume that larger dogs need a higher stimulation level. Dog trainers often find lower stimulation levels are great for dogs over 100 pounds. At the same time, your smaller dog might need a higher level. It's difficult to find the "sweet spot" for stimulation, but by paying attention to your dog's reaction and talking to a professional trainer, if necessary, it's 100% possible.
- **Vibrate or tone.** Find an e-collar that allows you to emit a collar vibration or audible tone. These are used as warning signs that the shock is going to happen. What typically happens is that your dog will start to notice the tone or vibration and stop the behavior immediately, so a shock isn't needed.
- **Distance.** It is helpful to get an e-collar with considerable

distance. A ¼ of a mile is minimum and a mile is usually maximum. The average length is ½ mile and perfect for house training. If you are more interested in hunting, you will find e-collars with one-mile distance.

Step One: Read the Instructions

Every e-collar should come with instructions. If you purchase a e-collar from a thrift store, garage sale, or a third party you can easily find instructions online. It is essential that you follow the instructions given to you. You always want to read the instructions before you put the e-collar together and attach it to your dog.

Step Two: Place the Batteries Into the Transmitter

Your transmitter is your remote. You will use this to control the e-collar, telling it when to give your dog a little shock. You want to ensure that the e-collar works with the transmitter and the levels aren't too high for your dog. You can do this by testing it on yourself first.

It is always best to put the e-collar on its lowest setting to start, even on larger dogs. Many people feel that larger dogs need a higher setting to feel the shock. This isn't necessarily true. Dogs are extremely sensitive to the shock they are given, no matter what their size. While a small dog will feel a stronger shock from a larger e-collar, bigger dogs can feel the same type of shock as a small dog from a smaller e-collar. Another reason to place the e-collar on the lowest setting is so you don't accidentally shock your dog when fitting the collar.

Step Three: Fit the E-Collar on Your Dog's Neck

With the e-collar turned off or on the lowest setting, fit the collar on your dog. Make sure that you can place your pointer and middle finger between the collar and your dog's neck. This is the general rule of thumb for any collar to ensure that it's not on too tight.

Be aware of e-collars that have small prongs. These have to touch the dog's neck, but you don't want to make it uncomfortable. You might need to adjust the e-collar a couple of times once you use it before you find the best fit.

One of the biggest mistakes is putting the e-collar too low. A dog's neck is shaped bigger at the base of their neck, causing the e-collar to move up the neck when they start playing. Here are some tips for fitting the e-collar:

- If your dog has thick fur, de-shed their neck first.
- Place the e-collar high on your dog's neck, but not at the top.
- Use two fingers to ensure it is a snug fit, but not too tight.
- Rotate the e-collar every couple of hours so your dog's neck doesn't get irritated by the collar.

Step Four: Let Your Dog Adjust to the E-Collar for a Week

While this step is optional, many professionals say it's best to not use the e-collar for a week. This will let your dog get used to the collar before it feels the first shock and not associate it with punishment. The point of the e-collar is to make the dog believe the negative behavior is causing the shock. If the dog knows it's the e-collar, the training can become difficult and they can try to escape from the collar.

Step Five: Start Using the E-Collar

Start with the lowest stimulation level and observe your dog's actions. If they react, such as moving their head or twitching their ears, keep the level low. If they don't react, upgrade the level. The dog should never whimper or try to run away because of a shock. If your dog makes noise like they are hurt, the shock is too strong and you need to lower the levels. If they're at the lowest level, you need to find a smaller e-collar for your dog.

Always be consistent when using the e-collar for training. For example, you don't want your dog jumping on the furniture, use the e-collar every time you see them do this. They might still do it when you aren't looking, but if you are consistent they will learn not to jump.

Step Six: Start with Commands Your Dog Understands

Your dog should already know basic training tricks, such as sitting. When first using the e-collar, start by saying commands they understand. If they don't respond, use the e-collar to get their attention. When the shock is received, repeat the command. Don't do this repeatedly over a few minutes as this can stress out your dog. Follow this step whenever you want your dog to sit, stay, or lay down. Every time your dog responds remember to praise them.

Step Seven: Control Your Dog's Negative Behavior

After your dog responds to basic commands, use the e-collar to control their negative behavior. For instance, if your dog jumps on company, use the e-collar every time you see them jump. While many people find their dog's growling or barking annoying, don't shock them immediately. Remember, this is how dogs communicate when something seems wrong to them. If they are barking because a stranger is coming toward your home, think about if shocking them is the best choice. Dogs are territorial and have natural protective instincts. You don't want to silence these instincts.

Do your best to ensure your dog doesn't see you using the transmitter. They are smart animals and will connect the shock to your transmitter. If they see you shocking them for their behavior, they will start distrusting you. You want your dog to believe the shock happens because of the negative behavior and not you.

E-Collar Safety

Don't Leave Your Dog Unattended with an E-Collar

Most people believe that you should have the e-collar on your dog, but this isn't necessarily safe. Your best step is to purchase the dummy e-collar and place this on your dog when you aren't training them, during the night, or when they are home alone. There is always a possibility that the e-collar will overcorrect or malfunction. While this is rare, it can harm your dog if it happens. Of course, because this is a rare occurrence, it is up to you.

Understand Your E-Collar

One of the best ways to practice safety with the e-collar is to ensure you understand everything about the e-collar. You have a vast amount of knowledge and know how to use your transmitter before you begin training. For example, you understand what all the buttons do and are prepared to follow your dog's reaction to know when the e-collar is at the right stimulation level.

Don't Use a Leash with the E-Collar

Using a leash with the e-collar can cause the front of the collar to push up against the dog's neck if the leash is pulled one way or another. If you shock them at the same time, this can harm them. You can use a harness that is attached to the leash, but don't attach any part of the leash or harness to the e-collar.

Don't Use the E-Collar When Your Dog is Swimming

While e-collars are waterproof, it's safer to keep them out of the water. If your dog is going to swim, remove the e-collar and don't replace it until their neck is dry. Following this rule not only secures your dog's safety but can make you e-collar last longer. When you need to wash the e-collar, always use warm and soapy water.

Benefits of an E-Collar

Fast Results

One of the challenges of training is that people don't have the time or patience to effectively train their dog. However, people who use the e-collar state it provides fast results and only takes a few shocks for the dog to stop the negative behavior. This works because before the dog feels the jolt, they will feel the warning vibration, associating this warning to the shock. Therefore, feeling the vibration makes them stop the behavior so you don't have to give them a shock.

Unfortunately, not everyone has had the quick success rates. Dogs that are stubborn tend to be harder to train. If you don't have an easily trainable canine, you shouldn't expect quick results.

Long Lasting Behavioral Changes

Professional trainers say e-collar training is the best way to change your dog's behavior, as long as it's used correctly. Studies have proved that dogs remember the behavior that brought on the shock better than a tone of voice or any other type of training technique. Hunting dogs trained not to go too close to the sheep received a shock. A year later, the dogs returned for another test and demonstrated hesitation when they started wandering too close to the sheep, proving they remember the shock when they got too close to the sheep. In fact, only one out of 114 dogs required a shock during the return test (Evans, 2018).

E-Collars Don't Take a Lot of Strength

When you use a leash for training, you need to have a tight grip and remain strong, especially with bigger dogs. They can easily pull a person around if you aren't strong enough to handle your dog. Strength isn't a concern with the e-collar because you use the transmitter to give the dog a warning noise or vibration and then a shock if their behavior doesn't change. Your dog can even be a ¼ of a mile away from you and still feel the warning and shock.

You Don't Stress Your Voice

Most dog owners know the stress your voice can feel when your dog doesn't listen to you, especially when their safety is concerned. For example, your dog escapes from their leash and runs into the street. Your automatic reaction is to yell at the top of your lungs for your dog to come back so they aren't hit by a driver. This can damage your vocal cords if you aren't careful or find yourself yelling often. With the e-collar, you don't need to say a word. You simply press the button and the e-collar does the rest.

Yelling or speaking to your dog angrily can cause them to become stressed, affecting them emotionally. They will become confused about what exactly they did wrong and worry about what will happen next. This poor communication between you and your dog can change with the e-collar. First, if you properly train them, they won't associate you to the e-collar. This will make them more comfortable with you. Second, your dog will receive clear communication in their current behavior by giving them a shock. They will come to understand that this behavior is unpleasant, making them stop.

E-Collars Allow for Easier Consistency

Training your dog with a leash or voice is difficult. Some people are afraid to train their dog in public using their voice for fear of causing a scene or judgment. The e-collar allows you to train your dog wherever you are without people realizing you are shocking your dog unless they see you push the button. If you do struggle with social anxiety and still worry what people will say or do when they learn you use an e-collar, you can purchase small transmitter to fit onto your belt or in your purse and push the button more discreetly. However, this should not become an issue because most people understand e-collars do not harm dogs.

Your Dog Receives Off-Leash Freedom

Leashes keep dogs in a certain area. While you can get a long leash, this can cause problems with your dog wrapping themselves around a tree or another object. Furthermore, if people aren't fully paying attention they can trip over the leash. E-collars will give your dog the freedom of running and playing without the tug of the leash. It is also safer as your dog can't tangle themselves up or trip anyone.

Common Myths About the E-Collar

The e-collar received a lot of criticism when it first became public. People automatically saw the e-collar as a harmful device for their companion. After all, who would want to shock their dog in order to teach them right from wrong? People who used e-collars were asked, "How would you feel if I shocked you when you did something wrong?" The harm an e-collar can cause a dog soon became a myth that still exists today.

The Shocks Harm Your Dog

Before I go too far into the world of e-collars, I want to tell you that the shocks your dog receives will not harm them. I love my dogs and have used e-collars for years. My dogs have never shown any signs of harm from the e-collars. Like you, I did a lot of research before I finally found the e-collar for my dogs. Since I purchased the e-collar, I will not go back to any other training technique. If my dogs ever showed signs of the collar hurting them, I would have stopped using it in a heartbeat. The shocks the dogs receive are similar to a static shock. While they are not fun to receive, they don't harm you.

If you take your dog to a training class that uses e-collars, they might have you try the e-collar yourself. I did this when I received the e-collar. I became surprised by the shock and felt more comfortable using it because it did not hurt me in any way. Therefore, I knew it wouldn't hurt my dogs.

Only Professional Trainers Can Use E-collars

Some people believe e-collars are difficult to use. While e-collars were difficult to use when they first came out, technology has changed. Today, they are easy to use. You don't need to be a professional trainer or talk to a professional to use e-collars.

The E-Collar Burns Your Dog

The e-collar does not burn the dog. There is no way that the e-collar can burn any dog because it does not get hot. The e-collar does not give off any type of heat when it gently shocks your dog.

The E-Collar Leaves Marks On the Dog's Neck

If you do see any marks from the e-collar, they are pressure marks. You have to rotate the e-collar on your dog at least every four hours. Furthermore, you should ensure you can place two fingers between your dog's collar and their neck. If you can't fit two fingers, the collar is too tight, making the e-collar uncomfortable for your companion.

Chapter 3: Your Dog and Their E-Collar

Before you choose an e-collar, you need to know why you want the training tool. For example, are you going to train your dog at home, for hunting, or to stop them from barking. The type of training goals you have depend on what e-collar you will get. At the same time, you need to remember your dog's size as not all e-collars fit well for smaller dogs.

Choosing the Best E-Collar

One of the toughest decisions you will make when e-collar training is what e-collar to buy for your dog. The list of e-collars in Chapter 2 should give you an idea of what to look for, depending on your training goals. For example, if you are going to house train, you want to focus on yard training e-collars.

The Fundamental Five

One way to help you choose the best e-collar is to follow these five tips.

1. *Choose an e-collar you can count on*

    Don't buy the cheapest e-collar for your training that you can find. You want to find an e-collar that is waterproof and has great reviews. Take time in looking for the best e-collar and check out online reviews on the product. Compare and contrast the reviews with different type of e-collars. This will help you narrow your decision a little more.

Book 2 - E-Collar Training Step-By-Step

2. *Education is always number one*

   Some of the best e-collars on the market will not only come with instructions but also training videos. You always want to focus on e-collars that show they are meant to train your dog and not punish your dog. Always remember, the e-collar is not a shortcut for training. It is a technique to help your dog understand that certain behavior is unwanted effectively and efficiently.

3. *Know your options when it comes to e-collars*

   Do as much research as you feel necessary to reach a complete understanding of e-collars. This means you want to research your dog's breed and how they handle training, know that your dog's age is going to affect training, and understand that some dogs are stubborn and will require longer training. This doesn't mean you aren't doing your job as their trainer, it's part of their personality. Look for an e-collar that holds constant and nick options and allows you to switch between the two easily.

4. *Choose an e-collar you understand*

   Some e-collars are more complicated to use than others. You don't want to pick an e-collar that will frustrate you. Pick an e-collar that doesn't have a lot of buttons as this means you need to look for the right button to press. By the time you find the button, the teaching moment is gone and your dog won't understand their previous action warranted the shock. They will relate the shock to their current action. Another method is to choose a small transmitter that you can fit in your pocket or clip to your belt for easy and quick access.

5. *Pick a collar only your dog can activate*

   There are e-collars that other dogs can activate, especially if you are using a no bark collar. When your dog receives a shock that wasn't meant for them at the time, they will become confused and unsure of why that happened, especially if they were listening, laying down, or sleeping. The best e-collars to purchase are the ones that will only set off a shock if it comes from the dog wearing the collar.

Your Dog's Reaction to the E-Collar

I have mentioned a bit about what to watch out for before, but I want to take this time to thoroughly explain your dog's reaction to the e-collar.

First, the e-collar can distress your dog. This is always a possibility and one that can happen even before you shock your dog for the first time. One of the best tips when it comes to knowing how your dog is going to react to the e-collar is to understand your dog before you even buy an e-collar. For example, if your dog came from a rescue shelter and is easily frightened, the possibility of your dog becoming scared by the e-collar is high. This isn't going to help the training situation, especially when they receive a shock.

Another way to know how your dog might react to the e-collar is to place a different collar on your dog. This is a good idea for dogs that don't have a collar. You will understand how they react to a regular collar and if you will need to speak to a trainer for the best advice on how to introduce your dog to an e-collar.

Unintended Reactions

It is always possible that your dog is going to have reactions you didn't expect to the e-collar. This can happen when they receive their first shock or throughout their training. You might believe you are training them not to leave the yard, when you are really training them to be afraid of someone. Let's look at a couple of case examples so you get a better idea of unintended reactions.

Book 2 - E-Collar Training Step-By-Step

Case Example #1: Tobey and the Neighbor

Tobey is a one-year-old poodle that likes to run over to the neighbor when he sees them outside. While this typically wouldn't be a problem for Tobey's owners, they are concerned for his safety because the neighbor lives across the street. The neighbor is always kind enough to walk Tobey back home, but anything can happen when Tobey sees the neighbor and runs into the street.

Other than the neighbor, Tobey doesn't leave the yard. Therefore, his owners weren't sure that the e-collar would be the best choice to train Tobey not to run out into the street, but they decided it is the best options for Tobey's safety.

Every time they bring Tobey outside, they put the e-collar on him. Every time Tobey ran toward the street because he saw the neighbor, he received a shock. One day, the neighbor came over and as soon as they walked in the door, Tobey bit the neighbor on the leg.

The neighbor and Tobey's owners were shocked by his reaction. They knew he loved to visit the neighbor, so why did he bite them? After speaking to Tobey's trainer, his owners found out that Tobey didn't associate the road to the shock. Instead, he associated the neighbor to the shock. This happened because Tobey's goal was to see the neighbor when he received the shocks. He wasn't thinking about the street or that this was the real cause of the shocks from the collar.

Case Example #2: Donnie and Max

Donnie's parents gave him a seven-month-old puppy named Max for his 10th birthday. The family loved Max, but Donnie's mother worried whenever he took Max for a walk. While she always went with, Donnie held Max's leash. Because Max is an energetic puppy, he often pulled Donnie, especially when he saw another dog.

After talking to her husband, Donnie's parents decided to purchase an e-collar for Max. On their walks, Donnie's mother would shock Max every time he started to pull Donnie. While she felt this worked well at first, she started to notice Max become anxious and aggressive whenever he saw another dog.

One day, Donnie and his mom were walking Max when they met family friends walking their dog. As Donnie and his mom started walking toward their friends, Max stopped and refused to get any closer. Because the family friends felt Max was tired or simply being a puppy, they continued to run up to Donnie and his mother. However, the closer they got, the harder Max tried to go the other way.

"I don't understand why Max started this behavior," Donnie's mother said to Max's trainer. "He always liked meeting dogs before. Now he becomes anxious, growls, and I am afraid he is going to attack another dog one day. What happened?"

"Max didn't associate the shock to pulling Donnie," Max's trainer stated. "He associated another dog with the shock. Therefore, Max believes that if another dog gets too close, he is going to receive a shock."

Dogs are not mind readers. They don't always understand our goals when it comes to their training, such as the case examples show. Both dogs would have received more efficient training without the shock collars. Because Tobey didn't run out into the road unless he saw the neighbor, his owners could have used positive reinforcement to sit on the sidewalk and wait for the neighbor to come over. The same goes for Max, positive reinforcement to not pull Donnie when another dog is spotted would have worked better. You always want to take time to think about what your dog's goals are when they are taking part in unwanted behavior. If they are trying to meet another dog or person, it is best not to shock them because they will associate their mission to the shock.

What If Your Dog Is Scared of the E-Collar

One of the challenges to e-collars you can have is your dog being afraid of the e-collar. This can happen due to two main reasons. First, the dog is frightened of any collar going around his neck. This can happen because of past experiences or because your dog is naturally more fearful than other dogs. Depending on the reason will depend on how you handle the situation. For example, a dog who is naturally fearful is going to overcome the fear easier than a dog who has a negative past experiences with a collar.

There is a lot of debate when it comes to using an e-collar on a fearful dog. Most people feel you should never use an e-collar on a dog that exhibits a lot of fear because their fear will increase. Other people believe this is one of the many myths about e-collars and if you train the dog correctly, you don't have to worry about their fear.

First, you need to understand where the fear comes from. If they are afraid of collars in general, you want them to get comfortable with a different collar and then work toward the e-collar. Even if you decide to leave the e-collar on for a week or two without delivering any shocks, you should start with a regular collar. Another tip is to introduce the regular collar slowly. For example, let them smell the collar and then sit it aside. Later, bring the collar back to them and set it down next to them. Watch your dog's reaction and if they start to show fear, take the collar away. Go slowly, step by step, until you can place the collar around your dog's neck without them trying to break free from it. Once they wear the collar for a week or two without trying to get rid of the collar, then work them into the e-collar. You might have to go a bit slower with introducing this collar because it is a different collar than the first one, but it will go smoothly after a while.

There is a possibility that your dog will not become comfortable with an e-collar. If your dog is from a shelter, a rescue, or from a pet store they may have terrible memories of collars. This trauma is something that they will never truly get over, just like any trauma a human can face. Always remember, there are other ways to train your dog that don't include the e-collar. Your dog's mental health is more important than any type of training.

Some dogs are fine when you put the e-collar on them and they get used to the collar. However, after you shock them once or twice, they become scared. For example, Robbie is a beagle that is learning how to hunt. His owner put an e-collar on him two weeks ago and has used it twice to get Robbie to come back once he has found the duck. One day, Robbie's owner tones him as a warning the shock is about to occur if he doesn't come back. Instead of heading back, Robbie drops to the ground and doesn't move. Concerned, his owner goes to him and notices he is scared. Robbie looks at his owner and whimpers a little.

A week later, Robbie and his owner are at a training class. His owner explains what happened and asked the trainer what to do. The trainer tell Robbie's owner that he is afraid of the shock. "This happens with some dogs. They understand the tone means a shock can happen and freeze because they don't want the shock and don't know how to protect themselves from it. Beagles aren't the easiest dogs to train with the e-collar for this reason, many tend to become afraid and freeze. Tonight, we can talk about other training methods instead of the e-collar. It's best that you don't use the e-collar unless you absolutely need to."

Your dog can have a lot of reactions to the e-collar. Some dogs are known to become more aggressive when they are shocked. If your dog has a reaction that shows fear or aggression, you should stop using the e-collar immediately. Your dog will respond better to a reward system and leash training. I would start by placing your dog into a training class and discussing any other training techniques with a professional.

The Story of Frankie and her E-Collar

Frankie is a black lab who was taken in by her owners, Amirah and

Thomas, because she couldn't become a show dog. She was born without any fur and while her first owners took her to the veterinarian and gave her medicine, fur never grew on the tip of her nose. Thomas, who worked with Frankie's initial owners, learned they were going to send Frankie to the shelter because they only keep show dogs to train and then sell. That's when Thomas said he would buy Frankie for his wife.

When Thomas brought Frankie home, she was seven months old. The couple had read up on how to raise a black lab because they knew she had natural hunting instincts, but they were not hunters.

Amirah and Thomas started the basic training with Frankie, but had a problem with her getting into the garbage. She would get into the garbage on the side of the garage, leaving it all over. Even when the couple caught her, they could barely pull her away from the garbage.

Worried about their dog's health from what she could eat in the garbage and tired of picking up trash all around their yard, they looked into other training methods. After researching the e-collar, they decided to give it a try.

Amirah, who works from home, spent her days training Frankie and followed the e-collar instructions. She let Frankie wear the collar for a week before she started to shock her. On the day, Amirah decided to start using the collar, Frankie walked toward the garbage. Once Frankie placed her two paws on the plastic garbage can, Amirah gave the warning and then a shock. Frankie moved her head a bit, but continued to go into the garbage. Cautiously, because Amirah didn't want to hurt Frankie, she gave the dog another warning and then a shock. This time Frankie barely moved.

Once Amirah got Frankie away from the garbage using her old tricks, she brought her dog in and adjusted to e-collar settings to a little higher shock. Later that night, Frankie got back into the garbage. Amirah gave her a warning and then shock, causing Frankie to jump back from the garbage.

When Amirah and Frankie came in, she told her husband what happened. They immediately called friends who use e-collars on their dogs and told them about it. "I would say that the level might be too strong for Frankie. She notices the shock at the first level you had her at, so I would move it back down. It sounds like you will need to give her the warning and shock before she gets to the garbage. Her mission isn't the garbage can, it is what is inside of the garbage can. Therefore, I don't think she will become frightened or aggressive toward garbage cans if she gets near one. Once she gets into the garbage, she is too interested in the garbage to care too much about the shock. Black labs are hunters and very determined dogs, it is hard to remove them from anything they set their minds on.

Amirah followed her friend's advice and noticed it helped. Within a couple of weeks, Frankie would walk toward the garbage, but then hesitate. The moment she started to hesitate, Amirah watched her dog closely. If Frankie took another step closer to the garbage, she gave the warning. If Frankie walked away, nothing happened. The dog wouldn't hear the warning sound. Soon, Frankie didn't worry about the garbage by the side of the garage.

Chapter 4: What You Need to Know Before Training Begins

Before you start training, you might feel overwhelmed about training your companion. At this point, it is important to remember that your dog and can sense your emotions. Therefore, if you are feeling anxious, they are going to start to feel the same way. They will associate their anxious feeling with training. This will interfere with their training and make their experience harder on you and your dog.

General Training Tips for Dogs at Any Age

There are tons of tips that you can use when training your dog. The most important factor to remember is you need to work with your dog. You need to do what is best for yourself and your dog. If you don't work together, the training system can easily fall apart. Below are several other training tips to help you and your dog thrive through the training experience.

Know Your House Rules

It's important that you don't just come up with a rule on the whim. Before you bring your puppy home and start to train, you need to make sure you know your house rules. You can do this by asking yourself basic questions. Do you want your dog laying on the couch? Are you going to let them freely eat or will you only feed them during certain times of the day? Are they going to have one area of the home they can stay in, run around the home, or will your dog remain outside?

Think about your dog's safety around your home. If you are getting a puppy, they can get into everything like toddlers do. It never hurts to ensure they can't open your cupboards with the cleaning products or get into your garbage.

No matter how hard you try, you won't come up with every single house rule right away. There are a lot of rules that you will think of once your dog comes into your home. For example, you might feel it's not okay to let them sleep with you on the bed, but have a change of heart once you bring them home. This is fine, I have done this myself, the key is you need to be consistent with your house rules. Don't change them once you have started training your dog because you "give up" on training or are "too tired to care." Inconsistency is going to confuse your dog and start to destroy trust.

Always Be Consistent

Consistency is one of the most important steps when you are training your dog. Consistency is one of the best ways they will learn. Furthermore, they will learn quickly, trust you, and know good vs bad behavior. Your dog wants to please you. Nearly every breed works on pleasing their owners. They don't want to do something wrong, but it is going to happen. It happens to everyone –human and animal–and is a part of the learning process. The more consistent you are, the easier your dog will adjust to their new home and rules.

Unfortunately, people struggle with consistency for many reasons. For example, you aren't home during the day, meaning you don't know what your dog does. When you come home and find they got into the garbage or tore up the pillow, you can't get after them at that moment because they won't understand. You always need to catch your animals in the act when teaching them what is right and wrong.

Some people aren't consistent because they have busy days and are tired when they come home. They tend to become relaxed with their dogs and let them do what they want because they don't want to train their companions. While I do understand how tired one can get at the end of the day, it is essential that you don't let yourself fall into this thinking trap. That's all it is—a thinking trap. When you decide to bring home a dog and train them, it is your responsibility to keep up on this training, no matter how tired you are. Don't relax on your training because the one who is going to suffer from that is your dog.

Here are some tips to help you focus on consistency in your training:

1. **Keep your daily routines**. Set a schedule before you bring your new family member home and stick to it as much as possible. Of course, you will have emergency situations and something might change here and there. But, the more consistent you are with the schedule, the easier consistency is with training. Another reason to keep your daily routine is because it teaches the dog that these are normal parts of their day. Have you ever taken a minute to think about how stressful life can be for a dog? They have a lot of situations that can cause them anxiety, such as walking in busy traffic or being home alone for eight hours a day. A routine will help your dog feel more comfortable about their day.

2. **Be consistent with your cues.** Your dog is about to get into the garbage can on Sunday morning, so you say in a stern voice, "No." Your dog backs away and goes to play. The next morning, your dog is going to do the same thing and you tell them, "No no" in a lighter voice. Your dog continues to go toward the garbage, which is when you say more firmly "No," and your dog backs away. The problem with this example is you are not consistent. If you say "no" in a firm voice once, you need to do that every time. Don't add another "no" or change your tone of voice as this will confuse your dog, which is why they kept going toward the garbage. The same goes with any nonverbal cues.

3. **Keep your words simple.** If you are training your dog to come to you by saying the following phrases, "come here," "come,"

and "come now," you are confusing your dog because they don't mean the same thing. Dogs listen to every word you tell them, and they will understand something different when they hear "here" and "now." Therefore, if you want your dog to come to you, simply use the word "come."

4. **Don't do all your daily training at once.** Training your dog for 20 minutes in one segment can become too much. They aren't going to remember everything, they will become confused, and they are going to get tired and annoyed. Your dog's attention span is similar to a toddlers–it is very short. The best option is to do any type of training in 2 - 3 minutes segments throughout the day.

5. **Everyone needs to be on the same page.** This can get a bit challenging if you have younger children who want to train, but everyone in the household should understand the house rules and train the dog in the same way. This can also be a challenge when you have company over. For example, you are training your dog not to jump on people, but your friend rewards the dog with petting and acting excited when your dog jumps on them. This will cause your training to take a step back. It is up to you to explain your training to your friend so they can help your dog understand that jumping on people is not acceptable. With everyone, including guests, being on the same page, your dog will quickly learn the house rules and stick with them.

Stay Healthy

Training is stressful, for both you and your dog. One of the best ways to give yourself the energy to train and keep your mind clear is to stay healthy. At the same time, you need to keep your dog healthy so they will have the best training experiences.

Making sure both you and your dog eat healthy is a great start to staying healthy. For dogs, you need to buy food for them that is nutritious, meaning you want to stick to their diet that is natural for them. This food will be a little more expensive than other food, but it will keep your dog healthy and strong. You want food that has a lot of protein, calcium, active enzymes, essential amino acids, and fatty acids. With these nutrients, your dog is more alert and pay more attention to what you are saying, your tone of voice, and your actions.

One factor you need to keep in mind with your dog's diet is transitioning. If you find you aren't feeding them the best balanced diet and find a different blend of food, such as Primal, you want to transition them to the new food. You want to start things off slow and steady, especially in the first week.

For the first two to three days, mix their regular food with the new food. At this point, you will have about ¾ their old food and ¼ cup Primal food. Watch your dog when they are eating to see how they react to the Primal food. If there are no problems and they eat it, add more Primal food and less old food on the fourth day by giving ½ old food and ½ Primal food.

It is possible that your dog can start having some bowel stress or gastrointestinal problems. When this happens, simply add one tablespoon of Goat Milk to their meal. Mix it well to ensure they get all the Goat Milk possible.

Once the seventh day rolls around, you can decrease the amount of old food to ¼ and increase the Primal formula to ¾. Then, on day ten, you will stop giving your dog any of their old and food strictly give them Primal food. While their system should be used to the Primal formula, it's always a good idea to monitor your dog's bowel movements for a few more days.

While you are working on your dog's diet, you can switch your diet as well. For instance, you might cut out sugars or eat food higher in healthy fats and lower in carbohydrates. Just like you do for your dog, introduce your new diet slowly and you will find that by day ten, both you and your dog are more alert and ready for the best training experience possible.

Getting enough sleep is another factor in staying healthy. It isn't always easy to make sure your dog gets the sleep they need, but they tend to do this pretty well. If you notice your dog struggling to sleep in their bed or in the area they are supposed to sleep in, do a little research to find out why. Another sign that your dog isn't sleeping well is they will find a different spot to sleep instead of lay in their bed. If you find your dog in a certain spot every morning, allow them to fall asleep in this spot and see if they remain their all night. Dogs will naturally get up throughout the night to stretch, check on you, or get some water. However, if they have a comfortable place to sleep, they will go straight back into that location until their day is supposed to start.

Be Patient

It is going to take time to train your dog and if they came from a shelter or have previous bad experiences, it will be harder. Dogs who come from abusive situations or are abandoned fear that these experiences are going to happen again. Like humans, they don't want to go through the physical and emotional pain it causes them. They want to feel loved in secure in their new home, but it is hard for them to trust you.

The best key to help your dog heal from any previous trauma is to be patient. Understand that your dog is emotionally and mentally hurting and you need to maintain a calm and trusting environment to help them through their emotions.

You also need to understand that some dogs will always have emotional and mental scars from previous trauma. They are similar to humans in this way, but aren't able to work through their emotions and mental scarring through therapy like humans can.

If you know or believe that your new family member was abused, here are some helpful tips to help your dog overcome their internal battles.

1. Get down to their level. Don't stand up and talk to your dog if they are afraid as this will make them feel inferior and increase their fear of you. Getting down to their level makes them feel equal to you.

2. Off them a treat. This isn't something everyone will do, but treats always make dogs feel better. Think about how often you want to have a treat after a bad day. Dogs feel the same way. Plus, giving them a treat reminds them that you love and care about them, immediately improving their mental state.

3. Make sure your dog has a safe place. You will start to know when your dog is uncomfortable or afraid through their reactions. When you notice your dog showing signs of this behavior, bring them to a secure place. This could be their area in the house, garage, or any place they feel the most comfortable. If you aren't in their secure place at home, take them out of the environment and spend quality time with them.

4. Don't forget about a pet behaviorist. It's important that you don't give up on your dog. Doing this will only increase their emotional and psychological trauma. If you find that you are having trouble handling your pet's behavior because of their past abuse or abandonment, bring them to a pet behaviorist. They will help you understand where your pet is coming from and give you ideas on how to help your pet overcome their trauma.

Reward Good Behavior

As a trainer, you don't want to get in the habit of shocking your dog with unwanted behavior. For success, you need to reward good behavior, including when your dog backs away from the unwanted behavior. For example, you are outside and training your dog not to go into the chicken coop. You have a curious puppy, and they often find the noises coming from the chicken coop interesting. Every time they pass the gate into the chicken coop, you give your dog a warning before a shock. Each time you see your dog, you ask them to "come" and wave your hand. They follow this direction and you give them a treat for coming to you when you called.

If you are training your dog and know you will need to give them a lot of treats, don't focus on unhealthy or bigger treats. Get some healthy treats that will help boost your dog and not drag him down. Too many unhealthy treats can cause stomach issues and make your dog sick. It's also possible to give him a little piece of a bigger healthy treat each time he follows your direction while training.

Obedience Classes

No matter what age your dog is, you can always enroll them in an obedience class. In fact, this is a great way to make certain your dog is socializing and you are receiving the right help you need for training. There is a lot of information that goes into training, and it is difficult to simply start training your dog without research and advice.

There are some dog owners who feel they don't need to spend the money on obedience training because their dog is "good enough." While you have a great dog, there are many benefits that are included in obedience training.

1. **You will meet like-minded dog owners.** While you may have friends with dogs, this doesn't mean that take training as seriously as you do. When you take your dog to an obedience school, you will meet people who have some of the same ideas and goals for their dog that you do. You will find someone who can help you through the training process or find someone to talk to about you and your dog's failures and successes with

training.

2. **You will expand your knowledge.** No matter how much research you do on training, obedience classes will expand your knowledge further. Your dog isn't the only one who will learn from the class.

3. **You will build your bond with your dog.** One of the best steps of obedience school is building your connection to your dog. This is a special time for you and your dog, and you will both feel it. Your dog will feel that you care about their general well-being and you will feel like you are doing everything you can to ensure your dog has a great training experience.

What Happens in Obedience Classes?

Sometimes dog owners are weary about obedience classes because they feel only dogs with behavioral problems go to them. This is a myth as any dog at any age will benefit from an obedience class. Furthermore, any dog owner will benefit from a class.

Another reason dog owners are cautious about obedience classes is they don't understand what happens in them. They feel it is simply teaching your dog tricks, but this is only a part of it. Obedience classes help your dog to understand what their role is with you and within the world. They will not only learn the basic commands, but also social skills. You will begin to understand healthy behaviors from negative behaviors when it comes to your dog. In general, you will feel closer to your companion because you will understand what they are telling you when they act a certain way.

Your Dog's Age Matters

The age of your dog matters when you are trying to train them. While dogs can be trained at any age, it is a lot easier to train puppies than older dogs. Some people feel they have to get a puppy if they want to train a dog, yet don't have the energy or ability to take care of an energetic puppy. Fortunately, this is a myth and senior dogs can learn just as well as puppies.

When it comes down to the basics, a senior dog is going to learn just like a puppy. However, a senior dog may take more time to learn the tricks than puppies. Part of this is because a senior dog is set in their ways, just like human adults get set in their ways. Another part is because senior dogs are generally slower. They are going to take more time to reach their paw up to shake or lay down. As long as you are patient and consistent, you will find your training fits with dogs of any age.

If you recently got an older dog from a shelter, get whatever information you can about their background as this will help you understand how your dog is going to react to training. Another reason is because you want to try to learn what tricks older dogs already know. For example, were they taught to sit by the previous owner and how? While this might be impossible to learn, you can work with your dog, a pet behaviorist, and a trainer to get an idea of how they were trained. You can start by telling the dog to sit–if they sit repeatedly at this command, they understand. If they don't sit continuously, they may have received inconsistent training.

For older dogs, you need to understand if they have physical disabilities that will stop them from doing certain tricks. First, you should never train a senior dog to do highly active tricks, such as learning how to surf or riding a skateboard. Senior dogs don't have the same strong bones as puppies and can get hurt easily. A veterinarian will tell you if your new dog has any type of disabilities and what they can learn vs what is too much for them.

Before you take in an older dog, you need to make certain you are ready for the responsibility. A lot of people think puppies take more responsibility, but in reality, senior dogs can because of any health problems. For example, taking in a blind dog with arthritis means medication and a lot of trips to the veterinarian's office. You will also need to care for the dog just as much as a puppy. For example, a blind dog will need to be slowly taken around their environment, so they don't constantly run into things. They will usually walk slower and be more cautious because they are afraid of running into something.

Case Example: Alice the Trainer and Her Dogs

I have raised many dogs throughout my life. My father bought my first dog, a Pitbull puppy, when I was eight years old. I named him Jake, and he taught me a lot about responsibility. When I was 15 years old, my parents allowed me to take in another puppy, a Black Lab named Buddy. To see the difference between Jack and Buddy amazed me. I knew that Buddy would be more energetic, but I never thought that Jack would become exhausted from Buddy. He really couldn't keep up with the puppy, no matter how hard he tried.

At one point, I noticed that Buddy wouldn't leave Jack alone. Even when Jack tried to go off to be alone and rest, Buddy tried to get him to play more or to get his attention in other ways. I quickly realized I needed to train Buddy not to bother Jack when he needed a break. At first, I thought of putting Buddy's e-collar on and shocking him when he would walk near Jack as he tried to rest. But I worried that this could cause Buddy to not want to play with Jack at all. Therefore, I started to train Buddy through a reward system. I didn't give him treats because I didn't want Jack to feel left out with the treats, but I would bring Buddy into another room and play with him or give him a treat without Jack knowing. Slowly, Buddy started to go to Jack when he went to lay down, sniff him, and then walk away.

This experience helped me realize the difference between young and old dogs, and I soon found myself focusing on professional training. Over the past 20 years, I have trained over 200 dogs of all ages. One of the first topics I discuss with all my clients is how the age of the dog matters when it comes to training, but only to a point. You can train older dogs, just like younger dogs, but you have to make sure older dogs get more breaks. They are going to get tired quickly. They are slower, but also determined to make your proud. They need the same consistency, patience, and care with their training that a puppy needs.

Training Your Puppy

Training a puppy is a fun, enjoyable, and stressful experience. While you enjoy seeing them learn and grow, they can also keep you on your toes. They don't have a strong attention span and can easily become distracted. However, they also want to do everything they can to please you, even the more stubborn breeds. Here are a few tips when it comes to training your puppy.

Your Puppy is Not an Infant and Not an Adult

Puppies are at a bit of an in-between age when it comes to the life of dogs. They are still growing and developing on a physical, mental, and emotional level. Yet, they are not infant dogs and can do more than you think by themselves. Because of their age, people struggle to know the best way to train their puppy. I know many trainers who have helped people through the difficulties of training a puppy.

Your Puppy's Developmental Stages

There are several developmental stages that your puppy is going to go through before they reach adulthood. You want to understand these stages so you can get the best out of them when it comes to training.

At four weeks, you should get your puppy socializing with other dogs. If you are a breeder or your dog had puppies, you shouldn't give them away until they are a little over eight weeks old. There is a lot of development that goes on within this time and allowing them to play with their brothers, sisters, and parents will help them grow mentally, physically, and emotionally.

Starting at five weeks old, they should interact more with humans. Of course, if you have children they are going to want to play with the puppies immediately. Try to hold off on letting them around the puppies too much until about week five.

At five weeks old, they will also start to become little investigators. They will get into everything they can and become interested in their environment. Let the roam around as much as possible, but also keep your eye on them. Puppies at this age can fit into the smallest places, causing them to get stuck.

Puppies need the first few weeks of their life to develop their dog-to-dog training skills. This is something you should not interrupt. By week eight, they will have developed more of their primary dog skills and might be ready for a new home. If you bring a puppy home around week eight, continue to focus on their social skills. Don't worry about full-on training yet and do not prepare an e-collar for your dog. Any dog should be at least six months old before you start thinking about an e-collar

By ten weeks, you want to make sure your puppy has had plenty of socializing opportunities and explored their surroundings. If they haven't they will start to become easily afraid of the unfamiliar. This is an anxiety that the dog will carry with them throughout their life, making them more difficult to train.

A Puppy's Fears

Before you start training, you have to understand your puppy's personality, including their fears. Most puppies are going to have some type of fears, especially in a new home. First, if you notice your puppy acting startled when they come across something or a person they don't recognize, this is natural. They will recover easily and try to see what this new stimuli is all about. They will continue to be curious and eventually grow out of the startling phase. However, all dogs will naturally be a little hesitant when it comes to new stimuli.

The trouble with puppies and their fears is if they don't show signs of recovery after a few minutes. There are many warning signs that will alert you to their fear:

- Trembling
- Scanning the room
- Lack bladder or bowel control
- They want to withdraw from the environment
- Whining
- They are trying to avoid what is scaring them
- Excessive panting
- Refusing to eat
- Vomiting
- Diarrhea
- Salivating

If your puppy starts showing some of these signs, of course they won't be potty trained yet, around five weeks old, they may have inherited fearful tendencies from a parent. At three months old, your dog will start to become anxious if you don't do what you can to help them through their intense fear. Once a dog becomes anxious, they will remain this way for the rest of their life.

Don't Start House Training Until Eight Weeks Old

While you shouldn't worry about an e-collar at this point, you can start basic training with your dog at eight weeks old. One of the first steps you will do is train them on their main spot. This is the area where they will eat, sleep, and spend the majority of their time. For example, if they will be an inside dog, they will remain indoors. If they are outside dogs, they go outside the majority of the time. It is essential that you teach them their potty spot at this age.

Training a Puppy Takes a Certain Mindset

Justin opened his own dog training business 25 years ago. The mission of his business is not just to help people learn how to train their dogs with an e-collar, but to overcome the challenges of training. "A lot of people tend to give up because they feel some forms of training are a lost cause. If they can't give their puppy to stop jumping on people, they become more relaxed about it because it's not worth the struggle. I have had people tell me they have other things to worry about than getting their dog to listen to them about everything. This is the wrong mindset to have when it comes to training. I always tell people who say something like this that they need to develop a more compassionate and tougher mindset. They need the compassion so their dog still feels comfortable around them. Compassion can also help them understand their dog needs consistency in training. A tougher mindset is more for them than their dog. This mindset doesn't allow you to give up. You want to remain strong because you want to do the best for your dog."

Training Your Older Dog

While most training tips are pretty general between puppies and older dogs, there are a few factors that are more important for older dogs.

You Can Train Them for a Longer Period of Time

Older dogs have a longer attention span than puppies. While you should start with shorter training sessions, such as 10 to 20 minutes twice a day, you can increase this amount of time little by little. For example, if you start with 10 minutes, two weeks in you can increase to 12 minutes. Always watch your dog and notice if they start to get distracted or too tired. You don't want to force them to go longer than they can handle.

Always remember, each breed is different. Some breeds are going to naturally have longer attention spans than other breeds. For example, Border Collies, Labradors, and German Shepherds have some of the longest attention spans because they have high levels of concentration.

You Need More Patience

You always need patience when training your furry friend, but you need more patience with older dogs. You can teach an old dog new tricks, but it is going to take longer than it will for a puppy. Puppies naturally catch onto new tricks quicker because of their age and curiosity level. Older dogs are not as curious, and like people, tend to slow down.

It's natural to feel like giving up on your training sometimes, you might even feel this way with a puppy. The key is to take a break if you need to–your dog might need a break too. Take a few deep breaths and think about all the process you and your dog have made. Sometimes people look so far into how much work they need to do that they forget about their progress. Don't let this happen to you as it will make you feel like training isn't working.

Socialize Your Dog

Just like puppies need socializing, adult dogs need to socialize just as well. No matter what age your dog is, they need to run, play, meet with other dogs, and people. Otherwise, dogs can suffer from fear, shyness, and loneliness. They won't know how to interact with people or other dogs. Other than the park or having your dog meet your friends and family, take them on walks with other dogs and dog training classes.

Get to Know Your Dog

If you don't know about your adult dog's past, take them to a dog behaviorist and a trainer. They can help you learn to understand your dog's past a little better by the way they act. Training dogs is always easier when you can understand certain behaviors.

Chapter 5: Let the Training Begin

Tips to Prepare Your Dog for Training

While I've already discussed several tips to prepare your dog for training, I want to bring up a few more. Some of these tips will focus on home training and some on obedience training.

Always Have Your Dog's Attention Prior to Training

E-collar training is not going to work if you don't have your dog's attention. When you first use the e-collar, you need a way to grab your dog's attention when you are going to give them a command. You might do this by tugging on their leash if you are outdoors or saying their name. You might decide to use some kind of motion, such as snapping your fingers twice. Once you can consistently gain your dog's attention, you can start with basic training.

Make Sure You Have a Little Play Time

If your dog is too energetic to focus on training, the whole process is going to fail. Play with your dog before you start training. You can take them for a walk or to the park. Do something that is outdoors as this will calm your dog out more than playing inside. Don't give them too much exercise because you don't want to make them too tired for training. The point is to let them run off enough energy so they feel calm and can focus on their training.

Have Everything You Need with You

You want all your training supplies right next to you when you start training. If you have to head to grab the treat, your dog is going to get distracted and then you need to start over. Place the treats in your pocket or hide them in some way so your dog doesn't notice them and focus on the treats instead of training. If you are going to implement the e-collar, have this available to you, if it's not already on your dog's neck. You can even have a few toys near you to reward your dog by playing with them for a while when they follow through with your command.

Empty Your Dog's Stomach

Don't feed your dog right before you are going to train them or head to a training session. If they need to use the bathroom, they will become distracted or could have an accident. While accidents can still happen, feeding your dog a few hours before you start training will help avoid accidents.

Every Dog Should Know the Basic Commands

There are a lot of tricks that you can teach your dog. You can teach them the basic commands or you can become more advanced. In general, what you decide to teach your dog is up to you and your dog. While every dog, no matter what age, should know the basic commands such as stay, sit and lay down, not every dog will want to learn other tricks. For example, your older dog shouldn't learn how to ride a skateboard. You don't need to train your dogs to jump rope, hide their head, wave, shake hands, or hug. You might feel these are fun tricks, but if your dog isn't into it, they shouldn't be forced.

Basic Commands

Finally, after all the information you've received it is finally time to start looking at some basic training strategies. The training techniques you will learn in this chapter don't all need an e-collar. While it is your choice to use an e-collar to teach your dog to sit or not, some professional trainers advise against it while others say it works great for basic training.

Some of the basic training techniques are best used with a reward system instead of a shock if your dog doesn't listen to your commands. However, if you struggle with training and feel you and your dog will benefit from the e-collar, then it is up to you. All you need to do is make sure your dog can handle the e-collar. For example, if your puppy is only four months old, even the smaller e-collars might be too much, but you can still try. You also need to remember to follow your dog. They will let you know if something is too much for them.

Bed Training

Some people feel that the e-collar should not be used for bed training. Like any training you do with the e-collar, it is generally up to you. If you feel you and your dog are ready for the e-collar, then use it for this type of training. You can also use this training to teach them to go into their create.

Step One: Facing the Bed

When you start this training, you want them to face their bed. Point to their bed and say "bed" and give a tone. Once your dog is all the way in their bed, turn your e-collar off. You should never have the e-collar on when they are sleeping. Always remember to praise your dog when they do something they should.

Your dog is going to be confused about the "bed" command at first; this is common when you just start training them in a new task. Don't spend too much time on forcing them to go to bed and don't stress them out by pushing the stimulation button over and over. You also need to be careful about disciplining your dog if they don't listen because they can associate this with bed. At the same time, if you give up you are telling your dog they have options and that's not building a strong training foundation.

You know your dog the best and you will probably come up with a solution to help your dog understand that the word "bed" means lay down and go to sleep. For example, you might find that putting a treat in their bed helps then understand or using hand gestures. Whatever extra training tool you use, you need to slowly reverse the action because you want them to go to bed when you give the command of "bed."

Step Two: Back Farther Away From the Bed

Once your dog is going to bed easily when you stand right next to his bed, you want to back up a bit. You can go into the next room and repeat the command. By this time, your dog will understand what "bed" and the tone means. They will more than likely head straight to their bed because you asked them to and they want the tone to stop. Dogs are great at figuring out how to get things to stop as they know what is in their best interest. Therefore, your dog is going to head to his bed because he hears the tone and the only way to get the tone to stop is by laying down in bed.

You can continue to back farther away from the bed as much as you want. For instance, if you allow your dog to roam all over the house, you never really know where you will be when you need to tell them "bed." Dogs tend to remember the scenery they are trained in and will associate this with the command. This does mean that your dog will understand the word "bed" and know what it means when you say it in the living room, but if you have never said it in this room before, they might not go right away.

Home Base and Perimeter Training

Every dog owner has the fear of their dog running off. Some people fear this so much that they don't let their dog outside unless they are with them and their dog is on a leash. Other people are a little more relaxed and will simply hope their dog stays within the yard.

One of the first basic training strategies that you need to teach your dog is their home base, sometimes called perimeter training. This is when you tell a dog where they sleep, where they can go in your house, where they can't go, where they will go potty, eat, etc. Dogs are not a companion you can walk around the house and say, "You go potty here, you go to bed here, you can't go in this room" and expect them to know. They have to be trained, and it will take time.

This training won't keep them at home. Your dog is going to run if they want and can run. The main point of this training is to help them understand what their role is in the home. It also establishes the owner/dog relationship.

Step One: What Are Your Dog's Boundaries?

To teach your dog where they can go and can't, you need to make sure you set up boundaries before training and stick to them. For example, your dog can go in the living room, but not on the furniture. They can't go into the kitchen, but can stay in the porch, walk-in coat closet, and entry way. You also decide your dog can go to the basement, but not into bedrooms.

You want to do the same thing with your yard. If you live in town, but don't have a fence, you will want to think about how to keep your dog within your yard. This might be difficult with a leash, but it is possible to do your best in training your dog not to leave your yard.

Step Two: Establish Home Base with Your Dog

Before you use the e-collar for home base training, you need to make sure your dog understands the type of training you will work on. You can do this by verbally telling them and spend a few days or a couple of weeks allowing your dog to become comfortable in the home, at least in the areas they are allowed in.

If you have used the e-collar before, you will understand that you want to get your dog comfortable before using. If you haven't used e-collar training previously, follow the steps in Chapter 2 when getting your dog used to home base training.

Step Three: Mark the Boundaries with Cones, Flags, or Anything Noticeable to a Dog

Your dog isn't going to understand the boundaries unless they are clearly marked. If you don't have a fence, sidewalk, or anything else to mark your dog's boundaries outside, you want to think of using cones or flags. You can then teach them that any place beyond the flag or cone is too far.

Step Four: Walk Your Dog to Their Home Base

No matter where the home base is, take your dog's leash and walk them to their home base. Once they are in the area, give them a treat. You will want to do this a few times throughout a series of days. Don't take 10 to 15 minutes and walk them to and from their home base and this may confuse them or stress them out. Plus, it would give them too many treats and they will become sick. Instead, take a couple of minutes to do this about two to three times a day.

Step Five: Reinforce with Verbal Commands

Some people will combine this step with step four right away. It is up to you to do this. I don't do this because I feel separating them gives your dog more time to understand their training. Remember, you need to have patience when you are training your dog.

You want to keep the commands short. For example, you might say "home" to let them know to go to their home base. If you have their bed at their home base, which is typically the case, you can say "bed." You probably won't want to use the verbal cue for "lay" or "lay down" as this should be saved for when you want to train your dog to lie down and relax around company or in another setting. Other words they will want to recognize when it comes to their home base is "stop" and "come." You will use stop when you don't want them to go any farther and come when you want them to come back.

Some people will also use a hand gesture during this phase. They may point to their dog's home base or move their hand in that direction. The biggest problem with using hand gestures is you can use the same gesture throughout your day and not notice. This will confuse your dog when you are sending them to their home base or trying to tell someone which direction to go.

Step Six: Repetitive Training

In this step, you will focus on perimeter training. This is when you look at the whole area your dog can go to and not just where their bed, food, or water is. When you are focusing on perimeter training, you want to make a distinction between home base and the rest of the boundaries. For instance, you will tell them to go "home," meaning home base, when you want them to go to bed or lay down. You will then walk them around the perimeter. You can allow them to walk freely as this will give you time to tell them "stop" or "come" when they go too far.

This step is going to take a while. You will want to spend at least a week, if not more, focusing on this step. Don't take your dog around too often when you are working on this step. While you want to repeat it for a while, you don't want to spend too much time training your dog where they can go and where they can't go. For a puppy, you should keep all training times to two to three minutes, an adult dog at about five minutes, and a senior dog can handle about 10 minutes.

At this point, you should only use the e-collar if you feel your dog is ready for it. Because this is still a basic training step, though at a deeper level, you don't need to use the e-collar yet.

Step Seven: Introduce the E-Collar

If you haven't done so yet, now is the best time to train your dog on their perimeter training with the e-collar. You want to ensure that they understand their training and are grasping the concept. You also want to make sure that you follow the right steps when introducing the e-collar.

Step Eight: Go Beyond the Perimeter

This next step is debatable for many dog owners as they feel it is a trick more than a test. In this step, you want to test your dog to make sure they do not follow you beyond the boundaries. Why many dog owners have problems with this step is because you need to ask your dog to "come" to you when you are outside of their boundaries.

The point of this step is to make sure they understand the importance of remaining in their boundaries. Another reason is because it helps set up the owner/dog relationship where you give them rules and they follow. When you use the e-collar, you will give them a warning and shock them if they pass beyond their boundaries. Of course, you can always use your verbal cues if you don't feel right testing your dog with the e-collar. Other people won't use the e-collar until they have reached this stage and their dog stays within the boundaries. It's always important to remember that dog training is a little flexible when it comes to you and your dog. You need to figure out what works for you and not what other people say.

Step Nine: Reinforcement Phase

In this step, you will allow your dog to freely move around the perimeter and remain hidden from their sight. However, you want to see where your dog is going. The reason why you want to watch is your dog is so you can give them a warning and shock if they cross the boundary. They won't know it is you and will think it is because they are going farther than the flags or cones allow them to.

You can also use this phase to test what your dog will do if something distracts them. For example, if they are outside, will they run toward a passing car? Of course, you will use your training method to bring them back into their perimeter. If they do become distracted, you know that you have more training to go before you can officially wean them off the e-collar for this training.

Sit Training

Whenever you start a new training technique, you want them to be in their home base area.

Step One: Say the Command

Use your attention grabbing technique to get your dog's attention. Say "sit" once with any hand gesture you are going to use to motion them to sit. This gesture needs to be different than the one you will use to tell them to lay down. When you get them to sit, reward them.

Step Two: Add in the E-collar

When you are training your dog, you can start using the e-collar right away, like we did with bed training, or you can wait until they understand the command. In this training technique, I am adding the e-collar in a bit later to show you how it works and why. You want your dog to understand their new command. You can make them feel stressed if you use the e-collar and they don't understand what they are supposed to do. This is going to get training off to a bad start and make you and your dog become too dependent on the e-collar.

When you add in the e-collar in the middle of training, you want the e-collar to replace your old techniques, such as a hand gesture or a treat. Your dog won't understand the change immediately, so you will want to use both techniques at the same time while slowly decreasing your old technique. For example, the first couple of commands with the e-collar, you will use both techniques. Then, when you feel your dog understands the tone or vibration of their e-collar with the command, you will stop using the other technique. If your dog responds, continue using only the e-collar. If your dog doesn't respond like they were, continue using your old technique at least two more times before repeating this process.

Once you say "sit" send the tone to the e-collar to get your dog associated with the tone (or vibration) and command. Remember to reward them if they sit. Your dog is going to make a mistake a few times and not sit when they are told to. This is normal for dogs, especially puppies, because they are easily distracted and still learning. Even if your dog doesn't listen to the command right away, when they do, follow it with positive reinforcement. Dogs thrive on positive reinforcement, and it will help them learn the command and want to perform.

Step Three: Practice

Just like you did with bed training, practice telling your dog to "sit" in various locations. You can even take them outside of the house to practice as you will have moments of telling your dog to sit wherever you go, such as a friend's house. The key when you practice is to not focus on the task too long and to remain consistent in your methods as much as possible. For instance, there might be times where you don't have the e-collar on and you need to tell your dog to sit.

Lay Down Training

Step One: Home Base

You need to make sure that your dog is in their home base area before you start the training. This is a consistent step whenever you start a new training method as it helps them learn. Plus, they are most comfortable in their home base environment. When your dog follows you into their home base environment, you will reward them. This is always an important step when it comes to your dog successfully completing any step in training.

Step Two: Say the Command

Tell your dog to "lay down." You don't want them to go to bed and lay down because this is going to confuse your dog with the "bed" and "lay down" command. You want to make sure that these two commands are separate and your dog understands this.

When you tell your dog to "lay down" you want to make sure you have their attention and you use a hand gesture or another type of technique they are used to in training. While you can start with the e-collar right way, like we did with the bed training, you can also wait to include the e-collar in the next step.

Step Three: Add in the E-Collar

This step follows the same format that telling your dog to sit does. You will add in the e-collar and turn it on right before you give the dog your command. Remember to reinforce the command with your previous gestures so they understand that the stimulation they feel is telling them to lie down. Again, even if your dog makes a few mistakes along the way, ensure that you always give them positive reinforcement when they follow the command.

Step Four: Practice

Once you are just using the e-collar to reinforce the command, start to practice in various areas around your home, outside, and even at a friend's house. Just as you did with teaching your dog to sit, you will use this command in various locations.

Another factor to remember when teaching your dog to lay down is that they will often follow commands and then get up back and want to play, especially puppies. You may have noticed this when you taught your dog to sit. It's important, especially with lay down, that the dog remains in that position for a few seconds, at least. You might want to train them to stay that way for a few minutes, but you always need to gradually train your dog. At the same time, you don't want to force your dog to lie down for too long, no matter where you are. This can cause them to become stressed, especially if they are energetic dogs. You should always think about your dog's personality when you are training them. This will help you ensure that you and your dog have the best training experience possible.

Come Training

Many people feel it is also important that you train your dog to "come" after you have trained them to "sit." This is because it will give them a sense of freedom from their last command. While some dogs might sit and then get up right away, especially if they notice you have a treat or toy in your hand, other dogs will sit for a period of time or until you ask move or train them to come to you.

Step One: Home Base

You always want to start training your dog at their home base no matter what you are trying to teach them. Most trainers feel that starting at the dog's home base is essential for this training.

Before you start training, choose a spot to stand within their home base. This spot is the area they will come to. If you move around too much during the first few times of the training, you will confuse your dog. They won't understand exactly where they are supposed to go when you say "come." You will focus on moving around later in the training.

Step Two: Say the Command

Following the same method you did with "sit" and "lay down," you will ask your dog to "come" while using a hand gesture or another method of training method they are used to. If you're comfortable starting training with the e-collar, it's fine to include it in this step.

There is a chance that your dog is going to come to you before you even give the command. They don't understand what is going on, so they are going to question why you are standing away from them. This is especially true for a puppy. Bring them back toward their bed or a bit away from you and try again.

When your dog performs the command, give them positive reinforcement. This doesn't always have to be a treat. Dogs love any type of positive reinforcement, including special attention and play time.

If you haven't used the e-collar to help reinforce this training yet, you should do so once your dog has performed the action a few times.

Step Three: Follow-up with an Additional Command

This command can easily make your dog wander, causing them to become confused. They may not understand when the command is done and continue to follow you. Because you always want to make sure your dog understands that the command is over, you can follow-up with another command. You can use any other trick that your dog knows, but most people use the "sit" command.

People will often use "sit" "lay down" and "come" training together. For example, you might ask your dog to "sit" before requesting them to "lay down." This is helpful for a dog because it gives them one step at a time. Some trainers will command their dog to do all three in a row. Of course, you want to do this slowly and make sure you are telling them on type of training at the time. For instance you will tell them to "sit" and then let them sit for a few seconds before commanding them to "lay down." Once they are down for a few seconds, you then command them to "come."

If you do this as practice, you want to make sure you don't place stress on your dog by performing the commands over and over within a small amount of time. For example, you don't want to spend a half hour repeatedly telling them to "sit," "lay down," and "come." You want to stick to a session of two to three minutes, no more than five minutes, for puppies and between 10 to 20 minutes for older dogs.

Stay Training

Step One: Home Base

Just like previous training, you want to start at your dog's home base. If they aren't at their home base, ask them to come to the home base. Make sure you give them positive reinforcement for following your command.

Step Two: Say the Command

You will start to teach your dog to stay with a hand gesture and your voice. Using the e-collar right away doesn't always work for this command at first because your dog isn't going to understand that they need to stay in that spot. Furthermore, many dogs will move around when they feel the tone, vibration, or shock from the e-collar. This will only cause problems within your training. You can include the e-collar after your dog understand what they are supposed to do when you say "stay."

Step Three: Reinforce with Another Command

"Stay" is a great command that is used after another command, such as sit or lay down. This is why most trainers believe you should teach your dog to lie down, sit, or come before you teach them to stay. For example, you command your dog to "sit." Once they perform this command, you tell them to "stay." This lets your dog know that they are not supposed to get up until they receive another command from you or they get tired of staying in the spot and want to move around. Always remember that you dog, especially puppies, need to move around regularly. They won't have the patience to stay in one spot for a long time. They need to get up and move.

Get Down Training

This training is a bit different to do because your dog needs to jump on something or someone before you can train them to get down. You can't really predict where this training will take place, so you don't have to bring them into their home base. But, you need to be prepared to start this training at any time.

Some trainers will cause the dog to jump up on something so they can start the training. You can do this if you want to train your dog in their home base, but they can become confused. For example, if they see you place the treats on the table and you allow them to jump up on the table, but then instruct them to get down they are going to receive mixed signals from you.

Step One: Say the Command

After you catch your dog jumping up on something or someone, command them to "get down." Use a hand signal or some other training technique to help them understand the command. You shouldn't use the e-collar at first for this command. They will need to understand what get down means before you use stimulation from the e-collar.

Step Two: Reinforce the Command with the E-Collar

Once your dog successfully follows the "get down" command a couple of times, add in the e-collar to reinforce the training. For instance, you tell your dog to "get down" but they don't listen, so you use the e-collar stimulation to reinforce the training. Like the previous training techniques, you will back away from the hand motion or other type of training reinforcement you used once you start using the e-collar.

Step Three: Everyone Should Positively Reinforce Your Dog

You know how important positive reinforcement is when your dog listens to a command, but the person they jumped on may not. Ask the person to give your dog some positive reinforcement for getting down when you commanded them to. You should explain this to everyone that your dog jumps on.

Chapter 6: Training Strategies, Levels, and Your Dog

By now, you know that your dog's breed and personality sets the tone for training, even with the e-collar. There are several breeds that are difficult to train, such as the Chow Chow, Akita, Siberian Husky, and Chinese Shar-Pei. You can train any breed of dog, but there are some that will give you a harder time than others. It's not because the dogs don't listen, it's typically because they have a hard time socializing or have a stubborn personality.

No matter what breed of dog you have or how old they are, you should always look into obedience school. These schools not only train your dog, but the trainers also help you with any problems you have or understanding training in general. It's another sense of support, especially when you and your furry friend are struggling with training.

Training Levels

Professional trainers of talk about how there are different levels of training. Depending on what trainer you talk to will depend on how many levels they focus on, but in general there are five. These levels might change depending on what school you go to. For example, parts of level one might make their way to level two. Some schools will often combine the first two levels because they are the easiest. This allows the dog trainers to focus more on the two harder levels. While you and your dog will go through these levels in obedience school, you can incorporate them at home as well.

Book 2 - E-Collar Training Step-By-Step

Level One: Foundation Training

The basic training we went through in the previous chapter focused on tricks your dog will learn in level one. This level is meant to teach the dog how to follow commands give them the basics to become a well-mannered companion. As the easiest level, most dog breeds catch on to this information and start to show signs of learning after the first class. Most dog trainers will not focus on the e-collar during level one. However, they may help you learn how to use it or talk about e-collar training at this level. Some schools will have special classes for e-collars at level one. You can always talk to your dog trainer if you are interested in bringing in the e-collar for training at this level.

Level Two: Skill Building Training

If the obedience school separates level one and level two, this level will build off the skills your dog already learned in level one. For example, the only trick you taught your dog in level one was to sit. This is because level two focuses on the exercises you will give your before they start training. A dog trainer in level two may focus on:

- Teaching your dog to "lay down"
- Teaching your dog to "come"
- Teaching your dog to sit whenever you stop walking

Level Three: Reinforce Reliability and Behaviors

By level three, your dog is ready for the more complex tricks. This is usually the level where e-collars are introduced and you can start to bring your dog's e-collar. Most obedience schools will allow you to enroll your dog in special classes, such as therapy dogs, mental exercises, and dog sports.

You and your dog might go through real-life situations where you will need to use basic commands. For example, your dog trainer has you and someone else act like you are walking your dogs down the sidewalk when one of them starts growling at the other. The trainer will then talk to you about how to handle this situation.

Level Four: Advanced Skills

You should understand that your dog doesn't have to take part in all of the levels. In a typical obedience school, you need to enroll your dog in each level. The only factor is you can't enroll your dog into level two without level one. You do have the follow the rules and prerequisites set up by the school.

It's also a good idea to make sure that you celebrate the accomplishments you and your dog have together throughout the training journey. Some people will celebrate with their dog, maybe by getting a special treat after each level. Your dog may not understand what the celebration is all about, but when you are excited your dog is bound to be excited. Furthermore, making it to level four is definitely something for you and your dog to celebrate!

Level four is the time when you start to show off all of the tricks your dog knows. Throughout this level, your dog will refine their skills, so they can amaze people even more with all their knowledge and tricks. At the same time, your dog will learn a few harder tricks or focus on sitting or lying down for a longer period of time. In this level, your dog may learn:

- Waiting at the door. This doesn't necessarily mean your door at home. They will learn to wait at any door. For example, if you head to the store and you can't bring your dog inside, you can leave them outside to wait. Of course, you always want to consider your dog's safety in this case. If you live in a small and rural area, it might be fine, especially if you know and trust everyone. You can also leave your dog in the care of another person you trust while you run into the store. However, if you live in a larger city, it is best not to leave your dog unattended at

all. Many dogs are stolen because they are left unattended in cars or by a store.
- Learning to ignore distractions and follow your commands.
- Walking with you while their leash is loose and ignoring any distractions, such as another dog or people.

Level Five: Expert Training Skills

For most obedience school, this is often the top level. Once you reach this level, your dog has an amazing ability to control their behaviors—most of the time. Remember, your dog is never going to be perfect at training. They will make a mistake from time to time, such as becoming distracted and not listening to your commands. When this happens, it is essential that you use your e-collar as it will motion the dog that this behavior is not acceptable. You should never use the e-collar to discipline your dog because they are not listening. As stated previously, your dog always needs to be exhibiting the unwanted behavior when you use the e-collar. Some of the behaviors your dog will work on during level five are:

- Listening to other people when told to go into a different position, such as when your dog is at the veterinarian's office.
- Staying in the "lay down" or "sit" for a longer period of time
- Sitting until you allow them to move
- Learning when they need to back up so they are not in the way

Practice Real Life Training

Professional trainers will talk to you about practicing real-life training moments before you are surprisingly put on the spot. This will allow you to think through the process you and your dog needs to go through to get the best outcome. When you practice training techniques this way, you can prepare yourself and imagine how you will handle certain situations. For example, if you bring your dog into a store (always make sure the store allows dogs) how will you handle your dog jumping on the shelf and knocking items down? Here are some examples of ways you can practice real-life training. These types of training experiences will work for any breed and dog of any age.

Practice Sitting Politely

One of the first tricks your dog will learn is to sit. One of the easiest training exercises you can bring wherever you go is teaching your dog to sit and wait for your next motion. This is going to be easier for some dogs than others. For example, if you have just started training your puppy, chances are they are not going to sit for a long period. They will sit and then get up a few seconds later. An older dog, especially a senior dog, will sit easier for a period, but it might be a good idea to motion him to lay down eventually. You always want to keep your dog's health in mind when you are training.

Whenever the opportunity arises to get your dog to practice sitting nicely–take it! It's a great way to get your dog used to sitting in various settings. Over time, you will notice him sitting for longer periods of time. This is a great perk about training consistently, your dog is going to adjust and they will find ways to keep themselves occupied when they need to sit or lay down because there is company over. But, no matter how patient your dog becomes, it's important to watch them and notice when they need to get up and move around. If your dog is well trained, such as they graduated from level five, they usually won't get up unless you motion them to. Don't let your dog sit there for too long that it becomes a bother for them.

Case Example: The Mailman and Alice

Alice is a young Miniature Schnauzer who loves to meet the mailman. Typically, this wouldn't be a problem, but Alice doesn't like the mailman to leave. She will try to bite his pants to get him to stay. Because Alice doesn't let go easily, she has started to ruin the mailman's uniform and he becomes a bit behind on his route. While he hasn't complained to Alice's owner, they don't want to cause any problems for the mailman. Therefore Alice's owner contacts a dog trainer they know.

The trainer told them to use Alice's best training skills to keep her from the mailman. "You don't want her to get too close, especially to his clothes. The first trick I would try is getting her to stay sitting when the mailman approaches. Also, keep the e-collar on for when she won't listen, but you need to be cautious about using the e-collar. You don't want her to become afraid of people, and this can happen if you send her the tone when she is heading to him. The best time to use the collar is when she doesn't listen to letting go of his clothing."

The next day, Alice and her owner waited for the mailman. The owner couldn't help but chuckle at her little furry friend because Alice stood so proud as she waited. Soon, she could notice him and became excited. "Sit" Alice's owner told her whenever she stood up. She would listen to the command, but once she became excited again, she would stand back up. By the time the mailman walked into the fenced yard, Alice couldn't contain her excitement anymore. She ran up to the mailman and followed him all the way to the house, which is their routine.

The problem doesn't begin until the mailman gets closer to the fence to leave. Right when Alice's owner noticed her dog about to go for his clothing, they snapped their fingers twice, catching Alice's attention. The owner then said "sit." Alice sat and looked at the mailman as he walked closer to the fence. Alice looked back to her owner, who didn't say or do anything. By the time Alice turned around, the mailman shut the fence door. He waved bye to Alice and carried on with his route.

Over the course of a couple weeks, Alice learned that she could follow the mailman to the house, but when he got close to the fence after dropping off the mail, she had to sit and allow him to leave. Because Alice is well-trained and her owners are consistent, the e-collar never had to be used for this situation.

Use "Come" When You Get a Chance

You don't need to be in a certain spot to tell your dog to "come." Once they are trained in this trick, you can use it when you want your dog to come to you. For example, if you want them to eat their evening meal, you will get their attention by calling their name and saying "come." The key is you need to listen to make sure they heard you. If they are playing in the next room with the kids, they might not hear you call them. Don't assume that you need to use the e-collar immediately. Take the remote and check to see if your dog can hear you call them. If you open the door to a lot of noise and they are playing, it's safe to assume they didn't hear you. Grab their attention and say "come." If they don't follow you because they are too distracted, then you will use the e-collar.

Take Your Dog for Car Rides

Most dogs love to ride in cars, so getting them into the car will not be a problem. If your dog is a little older, you may need to help them into the vehicle. Before you tell your dog to get into the car, you need to think of where you want your dog to stay. You also want to think of what you are going to do if you are driving and your dog gets in your way or a smaller dog tries to jump out of the window. Typically, these instances are rare and you can take precautions so they don't happen. For example, you will have your dog sit in the back seat until you know how they are going to react in the car. Another tip is not to leave your window low so that your dog can jump out.

Taking your dog on a car ride when you need to run an errand is a great way to practice "stay" with your dog. Of course, you need to make sure that your dog is safe and they won't get too hot in the car. Don't stay in the store long as your dog can become upset if they are not used to being alone in the car for an extended period of time. Remember to reward your dog with positive reinforcement when you return to the vehicle. If you have a larger dog, you can bring your dog out to play for a bit. They can stretch their legs or run around and catch a frisbee. It's always a great idea to bring toys with you just in case you have the opportunity to play with your dog.

There are a lot of drive thru places, such as banks, that will give your furry friend a treat and say hi. Even if it is through the window, your dog is interested in the new person and can't wait to get the treat they see!

Car rides are a great opportunity for your dog to learn more social skills. Other than the drive thru, you can take time to stop at a store that allows dogs and let them roam. You need to follow store policy, such as keeping them on a leash, but they will meet other people and possibly other dogs.

Case Example: Minnie and the Store

Minnie is a nine-month-old Poodle. Her companion, Paisley, adopted her from an animal shelter about two months ago. For help training Minnie, Paisley started taking her to obedience classes. Today, Minnie is in level three and struggling, so Paisley decided to take her dog out to a local pet store. She knows that Minnie suffers from a little social anxiety and wants her to become more at ease meeting new people and dogs. Paisley and Minnie's trainer talked about this situation, so Paisley feels prepared.

At first, everything went well. Paisley and Minnie walked up and down the dog isles where Minnie got to pick out a new toy. Paisley is now sitting on a bench where she is practicing "sit" and "stay" with Minnie when an energetic Golden Retriever comes walking up the aisle. At first, Minnie didn't seem too worried but showed interest in the dog. Minnie stood and started wagging her tail. Immediately, Paisley responded, "Minnie" to get her dog to look at her. Once she had Minnie's attention, Paisley told Minnie to "sit." Minnie obeyed, but still paid attention to the dog.

Looking at a dog book, Paisley's eyes scanned Minnie from time to time. She wasn't paying attention to the Golden Retriever until they became interested in Minnie. A young boy struggled to hold on to the leash of the Golden Retriever when the dog took off toward Minnie, causing the boy to let go. The boy started running and yelling at his dog, but Paisley remained calm. She got down to the level of the dogs and carefully watched Minnie and the other dog's reaction to each other. Both of them sniffed the other before they took a step back and looked at each other. Paisley become happy to notice that her dog hadn't moved, except for standing up.

Once the boy caught up to his dog, he grabbed the leash and apologized. Paisley told the little boy it is fine and talked about how the dogs are greeting each other. A few seconds later, the boy's mother came and the family left. Paisley looked at Minnie and praised her for being a good girl with the other dog.

A minute later, Minnie noticed another dog and started to show signs of distress. Paisley looked around and noticed a German Shepherd. This dog sat nicely by its owner and wasn't causing any trouble. Paisley started to care for her dog, trying to make Minnie feel better when the dog stood and started walking closer with its owner. Immediately, Minnie started barking and growling, causing the other dog to become defensive. Paisley started to practice the tricks that Minnie's trainer taught her, such as "no bark," but it didn't work. Minnie remained in defense mode.

Paisley started to feel a little panic, but remembered the trainer told her that Minnie will pick up on the panic, so Paisley remained calm. She then stepped in front of Minnie to create a block against the other dog. Once she noticed Minnie quiet down, Paisley turned around and said, "Minnie" and when her dog looked up, said, "walk" to motion Minnie to walk. Without a second thought about the other dog, Minnie turned and started walking toward the door.

The technique Paisley used with Minnie is known as "creating space" or "blocking the other dog." While this doesn't always work as some dogs will continue to look beyond their owner for the other dog, it worked for Minnie because they had trained for it in obedience school.

Aggressive Dogs

One of the most common problems dog owners have is meeting or noticing their dog becomes aggressive. It is important to note that your dog doesn't have to show signs of aggression prior to becoming aggressive. There are some situations, such as the above example with Minnie, when is it going to happen for various reasons. You may never understand why, but it is important to have a plan in place so you know how to handle these situations.

Paisley knew a bit about Minnie's history. She knew that Minnie lived with a larger dog that was mean to her at times. Therefore, Minnie would naturally become anxious or even aggressive when a larger dog approached her. So, before Paisley decided to take Minnie out to stores or around town where they could practice training in real life situations, she talked to Minnie's trainer, who helped Paisley come up with a few tricks to distract Minnie from larger dogs. Last on the list, was to send stimulation to Minnie through the e-collar. Because Minnie listened to Paisley when she said "go," Minnie did not receive the tone or shock.

Other than creating space, there are several other ways to handle aggressive dogs.

Keep Any Greetings Short

Some dogs don't mind running into another dog because they are curious animals. They want to smell the dog and say "hi," but then they want to continue on with their walk or don't care for the dog or the other owner. When this happens, your dog might become aggressive, just as Minnie did.

At the first sign you see of any type of aggressiveness, such as barking or growling, you need to take your dog out of the situation. One of the biggest mistakes people make is telling their dog "no" or to get angry. This is only going to cause the dog to become more aggressive because they feed off your emotions. At the same time, they might be warning you and they feel like you are ignoring this warning. It's always important to remember that dogs are protective. If they feel something isn't right they are going to take action. There comes a time when you need to listen to your dog and allow them to say "The conversation is over, let's move along."

In general, the best way to avoid any chances of this occurring is to keep greetings short. Even if you meet someone you know, if your dog is unfamiliar with the dog, person or seems uneasy, it is time to move on. Tell your friend you will give them a call and continue walking with your dog.

Another reason to keep greeting short is because your dog can become obsessed with the other dog. While this doesn't always pose a problem with aggression, it can for the other dog. If you do strike up a conversation, move your dog every few seconds and acknowledge them. This is also a good trick to follow if you meet a new dog and someone you don't know. Your dog can become stubborn and not want to continue on the walk because they want to say "hey" to the other dog. If it's okay with the other dog owner, let the dogs meet, but keep your dog moving at the same time. Slowly walk away from the dog and call your dog every few seconds to get their attention.

Notice How the Other Dog Is Acting

Even if you and your dog notice another dog approaching at the same time, you still have room to react to avoid a confrontation. Not only should you notice how your dog is reacting, but also the other dog. If you see any type of signs of aggressiveness from either dog, it's time to move your dog in another direction. You may have to do this by distracting your dog or creating a block from the other dog, so they decide to go across the street. Of course, the first method to try is simply walking your dog in another direction. Typically, dogs are very quick to follow their owners. If you do this before your dog becomes too interested in the other dog, you won't have a problem. One of the biggest keys to avoiding any aggressiveness is to stay one step ahead by noticing the signs. This is usually the best way to avoid any problems.

Avoid Other Dogs

Another way to handle meeting other dogs and dog owners is simply to avoid them. If you notice them, your dog is going to become more interested. If you don't pay attention to them, your dog is going to learn to just keep walking. While they might still look and try to get a sniff of the other dog, it is easier to tell them to come without much of a struggle.

Intermediate Level Tricks for Dogs

It is essential that all dogs receive basic training, discussed in the previous chapter. Now, I want to take training to another level and look at some of the tricks your dog can learn at the intermediate level. As long as your dog is healthy and interested in learning, they can perform these tricks. However, none of them are necessary. They are tricks that you can use as conversation starters or to surprise people. They are tricks that will allow your dog to show off.

Fetch

Teaching most dog breeds to fetch is pretty easy, it is almost like some of them have this trick in their body chemistry. But, there are other dogs that will struggle learning how to fetch. Not everyone feels the e-collar is necessary for fetching unless there is a possibility of your dog running away or causing trouble with people or other dogs.

There are three main skills your dog needs to know when it comes to learning how to play fetch:

1. Get it

2. Bring it

3. Drop it

Some people like to call the third skill "give it." The trouble with this is if your dog is playing with someone who isn't used to them, your dog could accidently bite the person's hand. For example, your nephews are visiting and want to play fetch with your dog. You agree and everyone heads outside. You've trained your dog to give you the ball, but you have to grab the ball out of their mouth for them to understand. You have never had a problem with your dog accidentally biting you or scratching with their teeth. Therefore, you aren't worried about anything happening to your nephews. In fact, you don't even think about the possibility.

As everyone is playing, you see your nephew hold on the ball in your dog's mouth. You notice your dog acting normal, but suddenly your nephew backs away and starts crying. You run to him, asking what is wrong. He shows you that one of your dog's teeth cut his finger.

You know this is an accident on your dog's part, so you don't discipline your dog. In fact, they look concerned about the situation. You take your nephew in, get the wound cleaned up well, and bandage it up. You know you have to watch it just in case there are signs of infection. You take a moment to think about how this can happen again as the neighborhood kids love to play fetch with your dog. It's at that moment you decide to teach your dog to "drop it" instead of "give it."

Talking to your dog's trainer, they tell you the best way to do this is by walking through the whole training process with them. Your trainer stated, "You don't want to just change part of the game without walking through the whole game. While your dog will have no trouble with the first two steps, you will need to take your time changing "give it" to "drop it."

One key tip for teaching your dog to play fetch is to have two toys that are exactly the same. You will play fetch with one while you hide the other one in your back pocket or somewhere out of their vision. You want them to focus on one toy at the time.

Step One: Tease Your Dog with the Toy

Whenever you want to play fetch with your dog, you need to get them interested in the toy. This is often caused "teasing." You can do this in many ways, as I am sure you have seen other people tease their dog before throwing a toy. You can talk to your dog to get them interested or act like you are throwing it in different directions before you really throw it.

Depending on your dog's age will depend on how excited they get for the toy and game. Older dogs might not be interested in playing fetch. However, younger dogs are going to love this game. If your dog isn't interested, try a different toy. If they are still not interested, they might be tired or want to do something else. Follow your dog's cues to notice if they are interested in something else. If a dog wants to play, they will let you know. If they are tired and want a break, they will inform you of this also.

Step Two: Toss the Toy

Once you get your dog's attention with the toy and they are ready to have it, toss it a few feet. Don't toss it very far at first as you still need to teach them the rest of the steps. When your dog picks up the toy and holds it in their mouth, praise them. You can run up to them and tell them what a good dog they are or use your regular praising method.

Step Three: Bring It

It is now time to teach your dog to bring the toy back to you. This can be one of the hardest parts because if a dog is not used to fetch, they will run off with the toy. Because of this, you want to make sure your dog understand where they can run off to and where they can't. If your dog goes beyond home base and you have their e-collar on, give them the tone to come back without saying or doing anything. They won't associate the tone or shock to their toy or play time, they will contribute it to going beyond their perimeter. Once they come back, praise them and continue to train them on bringing the toy back to you.

The easiest way to train your dog to bring the toy is to say your dog's name followed by "bring it." For example, "Princess, bring it" and when your dog takes two steps toward you, praise them. Go to them, grab the toy, and play a game of tug with them for a few seconds. This is when you gently try to take the toy out of your dog's mouth. The key is to let your dog win, meaning you let go of the toy while it is still in their mouth.

Step Four: Give It

At this point, you will give your dog the next cue to give the toy back to you. Take out the identical toy and tease them with it. This will get them to drop the toy in their mouth and focus on the toy in your hand.

Throw the toy a bit father this time, as you want to pick up the toy your dog dropped and place it behind you without them knowing you have it. Again, when your dog picks up the identical toy and walks toward you, praise them.

Step Five: Use the Word "Fetch"

You will repeat the first four steps by using the key words "get it," "bring it," and "drop it." After your dog has a handle on playing fetch, you will then start using the word "fetch."

When you throw the toy after teaching her, say "fetch, get it." Once your dog has the toy in their mouth, say "bring it," and when they bring it to you, say "drop it."

Once you start using the word fetch, you can slowly increase your distance. If you are at home, remember not to throw beyond your dog's perimeter. If they get distracted or go beyond their limits, use the e-collar to get them back into the perimeter.

Spin and Twist

When you teach a dog "spin and twist" you are teaching them 360 degree turns. This trick isn't typically taught to senior dogs as they don't have good balance. It is a great trick to teach puppies and some older dogs. Again, you know your dog the best. If you feel that this trick isn't for your dog, then you can skip to the next trick.

There are two main skills your dog will need to learn to accomplish this trick:

1. When your hand moves in the counterclockwise motion and you say "twist," they perform this action.

2. When your hand moves in a clockwise motion and you say "spin," they perform this action.

Step One: Halfway Spin

Start by having your dog stand or sit in front of you. With a treat between two fingers in your right hand, let the dog see the treat. Lure your dog in a clockwise half circle. Visually mark the point your dog reached and bring the treat back to you. Drop the treat by your right foot. This should get the dog to face you again. If they become distracted before or after the treat gain their attention. If this doesn't work, use the stimulation on their e-collar to gain their attention. Once they are calm and sitting or standing next to you again, move on to step two.

Step Two: Halfway Twist

Follow the directions from step one, but use your left hand instead of your right. Have your dog turn halfway counterclockwise and then drop the treat next to your left foot. Make sure you have their attention before moving on.

Step Three: Add the Verbal Cues

Once your dog has caught on to your luring, add in the verbal cues of "twist" when your dog moves counterclockwise or left, and "spin" for clockwise or right. Always remember to alternate between right and left.

Step Four: Start Phasing Out the Treats

Start with your right hand. Hold your hand up like you have a treat and continue to lure your dog half way. When your dog is looking at you, bring their attention to your left hand. Make sure they notice the treat so they don't become distracted by looking for the invisible treat. Once your dog makes the halfway turn, drop the treat.

Step Four: Complete the Circle

Keep using the treats in your left hand, but none in your right. Lure your dog through the whole way starting with a spin and then going for a twist. Remember to use your hand gestures and say the word as you are using the whole circle. Don't do this too much as your dog can become dizzy. Once they have follow command a few times, drop the treat and praise them.

Step Five: Add Speed and Mix Directions

Once your dog has a good handle on step four, you can start moving them a little faster and mixing up the direction. This means that you can have your dog twist twice, spin once, twist once, and then spin twice. At this point, you can keep the treat for when they are done spinning and twisting.

Chapter 7: **Advanced Training with E-Collars**

Now that you have the basic training and a couple of intermediate level tricks to teach your dog, let's focus on the more advanced training. When you get to this point with your e-collar, you will find that you don't need to use them too often. People will often send their dog the stimulation because they become distracted, aren't listening, or they leave their perimeter. But, because dogs are curious animals and easily distracted, it is always best to keep the e-collar on them when they are in the middle of training.

Will I Stop Training?

One of the biggest questions people ask if they will ever stop training their dog. My honest answer to this is it is really up to you. However, you should always practice the training your dog knows with them. While they will remember to "sit" when you tell them to, even if it's been a few days, if you become sloppy with training then they will follow this path. This means they will focus more on distractions than what you are saying. This will cause you to use the e-collar often. Plus, you will notice all the work you and your dog put into training is disappearing. So, while I do not advise that you stop training, it is always up to you.

Another reason you don't want to stop training is you can avoid having to reinforce any training. If you slack on training or stop telling them to sit, lay down, or stay they will start to ignore the commands. When this happens you will start to have trouble with your dog and become frustrated. You will tell them commands, but they will ignore you. This is when you need to reinforce all the training, starting with the basic tricks. To do this, you will want to start over.

Readjusting the E-Collar for a Growing Dog

Another factor to remember is to watch the e-collar. If your dog is growing, this means that collar has to grow with your dog. While you don't need to go through all the steps to get your dog adjusted to the e-collar, you will want to place the dummy collar on them when you need to clean the collar, charge it, or fix the strap so the e-collar will fit them correctly. Of course, if you don't always have the e-collar on your dog, you can do this when they are sleeping or taking a break.

Agility Training

To focus on agility training, you need a dog that is easy to train. Breeds that strive on obedience are the best dogs for this type of training. Dogs that are stubborn can learn agility training, but frustration can easily set in. Many people who have harder to train dogs give up when it comes to this type of training. It takes a lot of dedication and hard work from you and your dog, especially if they are a stubborn breed, but it is always worth it in the end.

Because of the in-depth process of agility training, I will not focus on how to teach your dog certain tricks. Most people will go to an obedience school to get the best agility training because of its advanced nature. Instead, I am going to give you tips to help you decide if agility training is right for your furry companion.

First, agility training is teaching your dog to run through obstacles. Dogs that perform on dog shows are taught in agility training. Some of the obstacles dogs usually learn are:

- The pause table
- Dog walk
- Tunnel
- Weave poles

Basic Training Before the E-Collar

Like with most training, the majority of people like to get past the basic training before they start using the e-collar. Instead, you will focus on a regular form of training you use, such as hand gestures, verbal cues, or a clicker. Clickers often work great because it is easy to gradually switch from a clicker to an e-collar.

Agility Training Is Great for Your Dog's Health

One of the biggest reasons people get their dogs into agility training is because it helps them stay healthy. Because they are so active, they burn off a lot of energy and they stay in shape. One of the most important factors is that you have to make sure your dog eats right or they will struggle when it comes to agility training.

There Are Risks

Unfortunately, agility training is a type of training that involves a lot of risks. Dogs need to pass a check-up by a veterinarian to make sure they are in good physical health for the tricks they will learn. This is an important part that you don't skip. For various reasons, dogs can have brittle bones and if they fall or trip over something while training, they could easily hurt themselves.

You also want to think about the heat. Your dog is going to run and stay highly active throughout their training, which tends to last longer than typical training. They are pushed by the trainers and will require more water, especially if it's hot outside. They will also need more breaks than normal if this is the case.

Using the E-Collar

When your dog becomes comfortable with the basic level of agility training, it's time to bring in the e-collar. Using the e-collar for agility training is different from other forms of training. For example, when you use the e-collar for home base training, you send the tone and shock to your dog when they leave their home base and don't come back when called. For agility training, you want them to keep running. You want them to run, jump, and follow the obstacles as much as possible. Instead of correcting your dog's behavior with the e-collar, you are redirecting their behavior.

When you redirect your dog, you are focusing on their agility route. They are following the rules and are running with you. For instance, Tammy's dog, Barnie, is learning his first agility route. As a puppy, Barnie is easily distracted and interested in almost anything he sees. Therefore, he tends to struggle following Tammy as she runs his route with him. When Tammy started using the e-collar, she decided should would call Barnie and ask him to "come." If he didn't listen, she would send him the signal from the e-collar. This immediately makes Barnie turn back to Tammy and they continue on his route. Within a week, Barnie is running the route beside Tammy without too much distraction. Within a couple of weeks, he is running the route by himself. Of course, Tammy still keeps a close eye on Barnie and will send him signals when he becomes distracted. Once Barnie reaches this point, Tammy and his trainer start adding new skills to his routine.

Adding New Skills

As the story about Barnie shows, you want to make sure your dog understands the basis of their route and know they need to keep going, ignoring distraction, before you add more skills. When you do get to the point of adding more, make sure that you only add one skill at the time. Once your dog incorporates this skill into their routine, then you can add another skill.

It is important that you continue to use the e-collar throughout this type of training. It is often known as the "silent enforcer" because you don't need to say anything. The second your dog goes off path, you send them the signals and watch them head back into their routine.

Roxie Jumped Over a Bar

Roxie's owner, Dustin, wanted to know if his dog would be good for agility training. After talking to Roxie's trainer, he received some tips and a guide to teach Roxie how to jump over a bar. The trainer stated if Roxie can accomplish this task, she is one step closer to agility training.

When it comes to agility competitions, dogs will jump as high as 24 inches. However a few have jumped higher. Dustin's goal isn't to get Roxie to jump high, so he decides to set the bar at its lowest setting, which is something the trainer advised him to do.

Step One: Walk Over the Bar

To get Roxie used to going over the bar, because all she wanted to do was go around it, Dustin got her leash and walked over the jump bar with her. At first, Roxie was a little hesitant to walk over the bar, but once she saw Dustin go over it, she followed.

Dustin repeated this action a couple of times and then started to say the word "over" each time they walked over the bar. Once they were over, Dustin gave Roxie positive reinforcement. Before he knew it, Roxie was interested in going over the bar again, specifically for the positive reinforcement.

Step Two: Roxie Walks Over the Bar Alone

Next, Dustin set Roxie in front of the bar and told her to "sit" and "stay." He then placed the bar on the next highest setting and said, "Roxie, come." The minute Roxie started to go over the bar, Dustin said "over." Dustin would repeat this action a couple of times before moving the bar up a notch.

Step Three: Roxie Jumps Over the Bar

Getting Roxie to jump over the bar wasn't the easiest part of the training, but Dustin understood this. Once the bar got to Roxie's elbow, she decided to dive underneath the bar. Dustin rewarded Roxie for her efforts and had her try again. This time, Dustin said "over" right as Roxie got to the bar. Roxie stopped, looked at the bar, and tried to jump over. Unfortunately, she knocked the bar down. Again, Dustin rewarded Roxie for her efforts.

It is at this point that Dustin starts to incorporate the e-collar. Instead of giving a visual cue, Dustin is going to say the word "over" when Roxie is supposed to jump. While she might not make it every time, if she refuses to jump, Dustin is going to give her a signal. Dustin's goal is to eventually stop saying the word "over" and simply give a signal to get Roxie to jump on time.

Step Four: Using the E-Collar

Dustin decided to help Roxie a little more by getting down to her level. Standing on his knees near the bar, he told Roxie to "come." Once she got to the bar, he told Roxie "over" and gave a signal. Roxie walked up to the bar, but simply looked at Dustin. He then repeated the signal. Right as Roxie jumped up, Dustin said "over" and gave another signal. He found Roxie on the other side. She had finally jumped over the bar without tripping or knocking it over. Dustin rewarded Roxie, and they continued this process a few more times without lifting the bar.

Boing

If you want to train your dog to jump over the bar but notice they struggle with their jump, then you can try a trick known as boing, which is when your dog jumps in the air on cue. When people use the e-collar for this type of training, they will usually switch to the e-collar instead of using a verbal cue. However, it is recommended that you start the training by using the verbal cue of "jump" or "boing" and not the e-collar.

Step One: Lure Your Dog to Jump

To lure your dog to jump, it's best to have a treat in your hand. You will crouch down a little and then spring up onto your toes. When you jump, you want to make sure that your arm is up as this is more likely to make your dog jump up because they want the treat.

Step Two: Use Verbal Cues

Once your dog starts to jump with you, add in the verbal cue "boing." Continue to jump as you were until you feel that your dog can jump up to your verbal cue.

Step Three: Use the E-Collar

When you start using the e-collar, you will still want to use the verbal cue as you send your dog a signal. Once you have repeated this action a few times, you can start to get your dog to jump higher. You will continue to use treats to accomplish this mission. For instance, you can first hold the treat between your fingers and raise your hand to where you want your dog to jump. Once they get high enough, you can hold the treat in the palm of your hand. By this point, you should have to use any verbal cues. Instead, you will focus on the e-collar's signal.

Hunting

Similar to agility training, you are going to use the e-collar for hunting once your dog understands the basics. It is also meant to redirect their behavior and not correct it. One of the basic tips for hunting with an e-collar is to get one specifically designed for hunting. While they are more expensive, you can reach your dog up to one mile, which is great when you want to call your dog back to you.

Teach Your Dog the Route

Before you take your dog out hunting, make sure that you planned out a route. You can start by taking your dog on a walk, preferably with a leash unless they are great at staying by your side and not running off. While you can use the e-collar, because you are training them for hunting, most people like to use the e-collar later in their training.

You want your dog to become used to the route. Take your time and let them smell and mark their territory. Don't hurry them along the route because they can easily forget it. You should take your dog on the route a few times before going onto the next step.

Make Sure to Train Your Dog to Return Home

Taking your dog hunting can be a scary time for a dog owner, especially at first. There is always a possibility your dog can get a bit lost–or you lost from your dog as they have a great smelling nose to find their way back to you or their home. However, you should always make sure that your dog understands how to return home. Before you start this process in your hunting training, you will want to make sure you and your dog understand home base training.

Another tip when it comes to training your dog to return home is to have a special signal. For example, you might give them two signals if you want them to return home instead of one.

Repetition Is Key

Like with all other types of training, repetition is key when it comes to hunting training. You always want to make sure you take time to show your dog what they need to do on their hunting adventure. Many owners will spend time every year retraining their dog for hunting, especially if they only go deer or duck hunting. This type of repetition is mainly for safety and to ensure your dog remembers what they are supposed to do.

Chapter 8: Common Mistakes

By now you understand that the e-collar is one of the biggest training tools for dogs. While there is controversy, if you use the e-collar correctly, your dog will thrive. Unfortunately, no matter how much you learn about e-collar training, mistakes can happen.

Lack of Consistency in Training

One of the most common mistakes is a lack of consistency in training. Because I have talked about consistency through this book, I won't spend too much time on it here. However, it is important to discuss as a mistake.

No matter where you are, if your dog does something that they aren't supposed to, you need to inform them of this correctly and immediately. Dogs are not going to understand what they did wrong 15 minutes ago. They live in the moment and need to be corrected right away. If you wait a few minutes, you are going to confuse your dog and they will associate the shock to whatever they were doing at the time. This can cause a lot of issues if you're not careful.

Training Your Dog for Too Long

It is important that you limit the amount of time you train your dog, especially in one setting. Younger dogs shouldn't be trained in one setting for more than five minutes while older dogs can generally last about ten minutes. If you struggle with time, you can set a timer on your phone. This will alert you to when it is time to give your dog a break.

When you don't pay attention to the time, you can overwork your dog. This can cause them to become stressed, tired, and sick over time. Plus, if it is hot you need to make sure that your dog gets plenty of breaks to cool down and water. They will need more water than normal if they are training outside on a hot day.

Dog Owners Don't Send a Signal Immediately

One of the most common mistakes happens when the transmitter isn't close to you and your dog misbehaves. For instance, you are in the kitchen when you notice your dog in the living room chewing on the couch cushion. You quickly scan the room for your remote and notice it is on the table in the living room. As you walk into the living room to grab it, your dog turns their attention onto a chew toy on the floor. Once you grab the remote, you push the button to give your dog a warning and shock. Unfortunately, your dog is no longer chewing on the cushion, so they associate the shock to their own toy.

If you don't keep your remote with you at all times, there will be moments you don't get to train your dog when they take part in unwanted behavior. This is something you want to avoid doing. Even if you are in your home, have the remote clipped to your pants or in your pocket so you can quickly grab it when you need to. This will not only make sure you can catch your dog when you need to, but will make training more consistent.

You Wait Too Long to Start Training

It is sometimes hard to know when the perfect time to start training is. There are general rules of thumb, but this doesn't mean it's right for every dog. One key is you want to start training your dog as soon as you bring them home. Even if they are potty trained and only eight weeks old, you need to make sure they start learning home base, where to sleep, where to eat, and the rules of the house right away. Many people don't think of these factors as training, but they are. Whenever you are teaching your dog something new, it is a form of training.

People Become Codependent on the E-Collar

The key to using the e-collar as a training tool is to get your dog to stop the unwanted behavior. Once the behavior as ceased, you need to wean your dog from the e-collar. If you continue to use the e-collar or you use it for all forms of training, you will find yourself become dependent on the e-collar. When this happens, your dog is also going to become dependent on the e-collar. Dogs who are dependent on the e-collar need to feel the warning vibration, tone, or shock to know they are doing something they shouldn't. If they don't feel one of these, they will continue to take part of that behavior.

You Don't Give Your Dog Enough Training Time

When you start training, you need to understand that you will train your dog often and you won't stop. Even when your dog has learned the basics, you will continue to train them. One of the biggest mistakes people make is they train their dogs the basic commands, when they want them to sit, lay down or get down, and then don't focus on any actual training time. This doesn't give your dog enough time to really learn the foundations of training.

Always spend time every day training your dog. Enroll them in an obedience class and get to know a professional dog trainer for extra help in case you ever need it.

The Dog Has Not Received Any Type of Prior Training

The e-collar is not meant to start training your dog the basics. It is meant to help your dog understand that certain issues are not appropriate, after they have received other types of training. For example, you will not use the e-collar on your dog when you are potty training them. First, most dogs are potty trained within a couple months of age, meaning they are too young for the e-collar. Second, using the e-collar when you catch your dog going to the bathroom can cause them to feel like that behavior is wrong. They won't associate the shock to the fact they didn't use the right are to go; they will associate it with their actions. This can make dogs feel that they are doing something wrong whenever they need to go potty.

Most people who use the e-collar right away, meaning without any prior training, are people who believe the e-collar is used as punishment or they don't understand the e-collar. You should never believe that the e-collar should be used as a form of punishment. Most e-collar companies advise against this and all dog trainers do.

You Use Harsh Discipline

It's a given that you will need to discipline your dog from time to time. For example, you might send them into a special kennel for a "time out" when they scratch the wall. However, most trainers state the best ways dogs learn is through training and positive reinforcement. While you may not completely believe this, you should never use any type of harsh discipline, such as hitting, staring down, grabbing them by the neck, jerking their leash, or yelling. Using this type of discipline can cause your dog to become aggressive and fearful. You can also harm your dog through some of these actions.

If you need help in understanding how to gain better control over your dog's behavior, the best place to go is obedience school. They will help you learn the ropes of training and make sure that your dog understands what behaviors are acceptable and which ones aren't.

People Don't Understand How the E-Collar Works

One of the most common mistakes people follow is getting the e-collar and immediately putting it on their dog and using it. Sometimes people will try it out, even if the dog isn't taking part in unwanted behavior, to see how their dog reacts. This is a huge mistake and something you should never do. You must allow your dog to get use to the e-collar before you use it. Furthermore, you have to wait until they are doing something you want to change before you give them a warning and then shock. If you are going to use the e-collar to train your dog, always use it correctly.

Chapter 9: Frequently Asked Questions and Answers

There are many common questions people ask about training their dog. Trainers find this part of their job enjoyable because they want to help people and dogs have the best training experience possible.

Question #1: Do Different Breeds of Dogs Learn Differently?

A lot of people wonder if they have a certain breed of dog, if they need to train them in a certain way. The straight answer to this question is no. All dogs, no matter what breed or age, learn the same. They can learn the same tricks with the same steps, tips, and strategies. As long as you are consistent and follow the steps within this guide, you can train any breed with an e-collar successfully.

Questions #2: Do I Wean My Dog Off the E-Collar?

Yes, you shouldn't continue to depend on the e-collar if you don't need to. If you feel your dog is successfully trained in the task, slowly wean them off the e-collar. You can do this by placing a dummy e-collar on your dog when you would have given them the e-collar. Remember, your dog shouldn't realize that the e-collar is what gave him the shocks. They should believe it was their behavior. As long as you trained your dog successfully with an e-collar, weaning is a breeze.

You can place the e-collar aside until you need to train your dog again or you find yourself bringing home another new family members.

Remember, when you are not going to use the e-collar for a period of time, you want to make sure it won't go off randomly, meaning turn your e-collar off and put it away.

Question #3: How Do I Know the E-Collar Isn't Harming My Dog?

The only way you will know is by watching and understanding your dog's reactions to the e-collar. If they whine, become frightened, cry, or show any signs of distress when you shock your dog, the stimulation levels are too high. Simply lower the level to the lowest setting and adjust as you need to. The dog should only slightly move its head when it feels the shock.

Most dogs respond to the lowest level of stimulation. While this does give them a little discomfort, it does not hurt them. Most trainers will tell you that a little discomfort is fine when it comes to keeping your dog safe and healthy.

At the same time, most trainers will tell you if you are afraid to use the e-collar, you shouldn't use it. As stated before, try the e-collar on yourself before you place it on the dog. This might ease your mind about how it will hurt your dog.

Question #4: How Do I Know When to Start Using the E-collar?

The simple answer is when you feel the time is right. Most people say that dogs should be at least six months old before they receive the e-collar because of the weight and size of the collar. The key is to do your research and make sure that you spend enough time going over your preparations for e-collar training. For example, you know your house rules, you know how you want to train your dog, and you are looking for a dog trainer to help you with the process.

Of course, you don't need to have a professional dog trainer on your side. However, it is always a great idea to get your furry friend enrolled in a training class for extra support and social time.

Question #5: I Have Seen Other Dogs React Negatively to the Shock from the E-Collar. How Do I Know My Dog Won't?

This is a great question. The truth is, a dog can react negatively to the e-collar for many reasons. First, they are not trained properly with the e-collar. For instance, the stimulation level may be too high or they receive a shock for every little thing they do wrong. You should never use the e-collar in this way. You only want to use the e-collar for certain training procedures.

Second, the dog may be more sensitive than other dogs. Just like humans, each dog has their own personality and there are several dogs that are highly sensitive. This will cause a dog to feel the shocks strongly, stronger than other dogs. This will also cause them to internalize the shock more than other dogs, causing them to become sad or even depressed. In general, if you have a highly sensitive dog, you might not need to use the e-collar. Highly sensitive dogs tend to follow instructions through a person's tone of voice a lot better than anything else.

Conclusion

If your head seems like it is overloaded with information, this is typical when it comes to dog training, especially if you are beginning this journey. Fortunately, this book is here for you whenever you need it as you can simply download the book and have it in your phone or on your device for whenever you need it.

Even if you feel you might need to reread some information before you start training your dog, especially when it comes to the training steps, I know you have a better idea about e-collar training and how you can incorporate the e-collar into various tricks.

One great fact about this book is it not only explains what e-collar training is, but you also receive a number of tricks to teach your dog. This book didn't simply focus on the basics, it also looked at more advanced training techniques.

Some of the most important takeaways from this book are:

- Making sure you dog becomes used to the e-collar before you bring it into training. You never want to put the collar on and start giving your dog signals. This is going to cause your dog stress as they will be confused about what is going on. Furthermore, they will know that you are doing this to them and the trust between you and your dog will start to fade.
- You have to be consistent as a dog trainer. It doesn't matter if you are in your home or at a friend's house, when your dog does something they shouldn't, you need to be quick and correct their behavior. Remember, dogs only focus on what is going on at the moment. If you give them signals from the e-collar a few seconds

after the unwanted behavior, the dog is going to associate the signals to what they are doing at that moment. Always notice what your dog is doing when you are in the middle of training, and be ready to act to help them learn that some behavior is not acceptable.

- E-collars will not harm your dog unless you have the collar on too tight or the stimulation level too high. Always make sure you can fit two fingers between your dog's neck and the collar. Start the stimulation level at the lowest setting and wait until you need to use it on your dog to correct their behavior. If they turn their head or react in a slight way, they noticed the stimulation. If they didn't respond at all, it is time to adjust the level. Remember to only move the stimulation level up one. If your dog whines or whimpers when you give a signal, the level is too high and you need to bring it back down.
- Make sure that you don't buy the wrong collar for your dog. Some e-collars are not meant for small dogs. Dogs that are under a few months old shouldn't be given an e-collar. While most people feel the general rule is about six months old, you can start your dog on the e-collar earlier. It really depends on the amount of training they have. It is important to note that there are tons of great e-collars available to you. There are so many that I couldn't include them all in this book. Do your research to make sure you really do pick the best e-collar for your dog.
- Always find a way to gain your dog's attention before you start training them. Even if you've trained your dog for two years, you always want to grab their attention. Dogs are naturally easily distracted, and it isn't fair to them to send them signals when you didn't let them know that you wanted them to follow your command. Use their name or find another method, such as snapping your fingers twice, to gain their attention.
- Always reward your dog with positive reinforcement when they are training. Even if they make a mistake, you want to reward their efforts. You don't always need to give them a treat. Taking

time to play with them and talking to them is a great way to give your dog positive reinforcement.
- While dogs can receive training at any age, you always want to keep your dog's age in mind because at the end of the day, it does matter. Puppies are going to be more energetic, which can make you feel more frustrated with training. Older dogs are going to be slower, which means you need more patience. Another factor about older dogs is while they can train for longer periods of time, they will tire out sooner. Never push your dog to finish training or train longer when they are showing your signs that they are tired.
- Start training your dog as soon as you bring them home. Even if they are only 10 weeks old and are potty trained, you want to focus on the basics, such as where they eat, sleep, and their home base.
- Do whatever you can to make sure your dog eats healthy and gets enough sleep, as much as you can do this for a dog. Eating healthy and getting enough sleep will help ensure that you and your dog have the best training experience possible. Furthermore, this is something to do with your pets whether you train them or not. It is up to us to make sure they get food with the nutrients they need so they can live a long and healthy life.
- Don't become codependent on the e-collar. While you will use it often, most people believe the point of the e-collar is to correct your dog's behavior and then wean them off. If you become too dependent on the e-collar, then your dog will only react when they receive the stimulation. You want your dog to react to your commands and not the stimulation of the e-collar. Always remember the e-collar is more of a reinforcer when it comes to training.
- Dog training takes a lot of patience. You will need to repeat the steps over and over to give your dog the best training experience. Always remember that the number one factor with training is the safety and health of your dog. Once you keep this in mind, you

can give yourself a breather when you start to feel frustrated or losing patience. You are trying to help your dog and not harm them in any way. Rushing through training is going to harm them mentally and emotionally. Furthermore, if you have the e-collar on too high and find yourself giving them signals often, you could end up physically hurting them.

- Follow the directions on how to put on the e-collar to ensure it is in the right location. If you place the e-collar on the base of your dog's neck, it will move up the neck when your dog is playing, This can cause them problems and they won't feel the signals.
- Enroll your dog in an obedience class. This will not only help your dog when it comes to training, but it will help you as well. You will meet a great dog trainer who can help you through some of the toughest times and meet like-minded individuals. Everyone you come in contact with in obedience class will become a part of your journey. Furthermore, your dog will get some social time, which is important. If you only have one dog, your dog can become lonely and feel isolated if you don't allow them to play with other dogs and meet other people. Obedience training is one of the best ways to do this.

Above all, you need to enjoy your time with your dog. Do what you can to make dog training fun. Of course, it will be more work that you can imagine. However, as you work with your dog and listen to them, they will listen to you. If you take the advice given to you about how to train your dog in basic commands as well as more advanced command, you can watch your dog thrive with everything he learns. Through the tips given to you in this book and remembering the key takeaway notes, you will have the best training journey possible with your furry companion. This will give them some of the best memories throughout their life. You will have a great relationship and your dog will spend many years as one happy, obedient, and loving dog.

Paul Davis

References

5 Tips for Dog Training Session Prep. (2017). Retrieved 1 October 2019, from https://acmecanine.com/5-tips-dog-training-session-prep/

Accessories & Replacement Parts | E-Collar Technologies. Retrieved 28 September 2019, from https://www.educatorcollars.com/accessories

APDT - "Real Life" Training. Retrieved 2 October 2019, from http://www.trainyourdogmonth.com/tips/reallife.aspx

Arterburn, J., & Benson, K. Dog Shock Collar Myths and Misconceptions. Retrieved 21 September 2019, from https://www.securepets.com/debunkingmyths.html

Barkless Dog Collar | Barkless Pro Anit Bark Collar - E-Collar Technologies. Retrieved 26 September 2019, from https://www.ecollar.com/products/barkless-pro-anti-bark-collar

Bender, A. (2019). Our Top 10 Puppy Training Tips. Retrieved 29 September 2019, from https://www.thesprucepets.com/top-puppy-training-tips-1118511

Bennett, J. 7 Must-Read Tips When Buying a Purebred Dog. Retrieved 26 September 2019, from https://www.rover.com/blog/purebred-dog-tips/

Choosing a Puppy. (2019). Retrieved 17 September 2019, from http://www.apdt.co.uk/dog-owners/choosing-a-puppy

dog agility training for beginners: how to get started - SitStay. Retrieved 2 October 2019, from https://sitstay.com/blogs/good-dog-blog/dog-agility-training-for-beginners

E-Collar for Dogs - Remote Training Collars | E-Collar Technologies. Retrieved 27 September 2019, from https://www.ecollar.com/categories/remote-dog-trainers

Evans, B. (2018). Benefits of using Dog Training Collars that you can't ignore. Retrieved 27 September 2019, from https://petspy.com/blogs/dog-training/benefits-of-using-the-e-collar-training-that-you-can-t-ignore

Finlay, K. 5 Reasons You & Your Dog Should Take An Obedience Class. Retrieved 29 September 2019, from https://iheartdogs.com/5-reasons-you-your-dog-should-take-an-obedience-class/

How to Care for a Formerly Abused Pet. (2016). Retrieved 29 September 2019, from https://www.tasteofthewildpetfood.com/training-behavior/how-to-care-for-a-formerly-abused-pet/

Howell, A. (2019). The E-Collar Dog Training Bible : The All-Inclusive Guide, Including Specific E Collar Training For Golden Retrievers, German Shepherds, Labrador Retrievers, And Beagles. Kindle Edition.

Hunting Dog Collars | E-Collars for Hunting Dogs - E-Collar Technologies. Retrieved 26 September 2019, from https://www.ecollar.com/categories/hunting-dog-trainer

Krohn, L. (2017). Everything you need to know about E Collar Training. Retrieved 2 October 2019, from

Obedience Courses. Retrieved 2 October 2019, from https://www.animalhumanesociety.org/behavior/obedience-courses

Pavia, A. (2017). Adopting A Shelter Dog: 5 Tips For Success. Retrieved 23 September 2019, from https://fearfreehappyhomes.com/adopting-shelter-dog-5-tips-success/

Puppy Training and Socialization Tips for Owners. (2011). Retrieved 29 September 2019, from https://healthypets.mercola.com/sites/healthypets/archive/2011/09/22/every-puppy-owner-must-know-about-early-training-socialization.aspx

Ray. (2010). History of the Shock Collar. Retrieved 28 September 2019, from http://dogtrainingclub01.blogspot.com/2010/12/history-of-shock-collar_20.html

shibashake. Dog to Dog Aggression – Why and How to Stop It. Retrieved 2 October 2019, from https://shibashake.com/dog/dog-to-dog-aggression

Stregowski, J. (2019). Are You Guilty of These Dog Training Mistakes?. Retrieved 2 October 2019, from https://www.thesprucepets.com/common-dog-training-mistakes-4030442

Things to Consider When Choosing a Dog. (2017). Retrieved 17 September 2019, from https://trupanion.com/blog/2017/03/choosing-dog/

Train your dog: The relevance of consistency! | Tractive. (2018). Retrieved 28 September 2019, from https://tractive.com/blog/en/training-en/consistency-and-rituals-in-dog-training

Transitioning for Canines. Retrieved 29 September 2019, from https://primalpetfoods.com/pages/transitioning-for-canines

Where to get a puppy. Retrieved 28 September 2019, from https://www.humanesociety.org/resources/where-get-puppy

Wildesen, A. The Misunderstood Doberman Pinscher. Retrieved 17 September 2019, from https://thecaninetrainingcenter.com/the-misunderstood-doberman-pinscher/

Wilson, S. (2019). 8 Things You Need To Know Before Buying A Shock Collar | CanineJournal.com. Retrieved 27 September 2019, from https://www.caninejournal.com/shock-collar-for-dogs/

BOOK 3

TRAINING YOUR PUPPY STEP-BY-STEP

A How-To Guide to Early and Positively Train Your Dog. Tips and Tricks and Effective Techniques for Different Kinds of Dogs

PAUL DAVIS

Book 3 - Training Your Puppy Step-By-Step

Introduction

One of the greatest joys in life can be the little pets that we have at home. Watching a dog grow from a teen puppy to a well-trained furry companion is an exciting thing. Many individuals believe that having a puppy is going to be a fun experience. It certainly is! At the same time, the hard parts of training are often overlooked. Some people can get in over their heads and end up with too much to handle when trying to properly train their four-legged friends.

You might be the most responsible pet owner out there. You could have only fresh organic treats and food, the top-quality grooming and veterinary care, and a personal trainer for your dog. No matter what you do, there are no circumstances that will keep you free from the slight chance that you might wake up in the morning to find a trail of diarrhea across the house, an expensive shoe that's been ripped up, bits of a dollar bill found in vomit, a school paper soaked in pee or a wad of toilet paper dug up from the trash can and ripped to bits strewn about a house.

Dogs are cute, but they are animals, which means they aren't free from having their nightmarishly disgusting moments. They might get fleas, ticks, and other creepy critters that can make your stomach turn. This isn't to turn you off from getting a dog! Nor are these aren't daily experiences. However, they are occurrences often overlooked by those wanting to adopt a pet, which is why millions of household pets end up in shelters every year. Don't think that having a dog is all disgusting, but don't think that you'll be free from these random instances.

Nothing we discuss is meant to scare you away or make you nervous about having a dog! With all the animals on the streets, in shelters or even in abusive and neglected homes, it is important to know everything that you are getting yourself into before adopting a pet. These are innocent animals who are going to look up to you from the moment you bring them home.

Don't think of your puppy as a human. If anything, think of it like a human baby, but throughout its entire life! They will be smart enough to learn new tricks, but they will also not realize that eating garbage is wrong. Throughout this book, we will give you a comprehensive guide of everything you need to know so you can successfully raise a healthy pup.

There are 8 steps to training your puppy:

1. Pick out your dog.
2. Prepare your home.
3. Start basic training.
4. Socialize them and prepare them for the world.
5. Teach them more challenging skills.
6. Ensure there is proper exercise.
7. Keep up with the health of your dog.
8. Avoid the mistakes and reflect on what you can improve.

Throughout each chapter, we will give you all the info necessary to help you create your training plan. In the final chapter, we will discuss how you can create a specific schedule along with other important reminders to live by throughout this process.

It is an exciting time and you can be assured once you bring your furry friend home, life will never be the same! Let's get started now in discussing how to choose the right dog for your home and your lifestyle.

Chapter 1: Choosing the Right Dog

Most people give up their animals because they can't afford them (Greenwood, 2016). At the same time, many individuals will end up giving their dogs away because they simply can't keep up with their basic needs.

Picking the right dog is important. There are a few options that you have to get started. You could go to a shelter, a breeder or a pet store. Once you have decided where to get your dog from, do your research on the specific breed. Throughout this chapter, we will give you tips on how to best choose the right dog, along with a few suggestions for people with specific needs. If you don't get a dog that matches your lifestyle, it can make training them more challenging, while weakening the relationship.

Consider all factors that we discuss in this chapter when choosing the new family member you will be adding to your life.

Picking Where Your Dog Comes From

With over 900 million estimated dogs in the world, we can be certain there won't be a shortage anytime soon (World Atlas, n.d.). When choosing to bring a new dog into your home, you'll have several different options to choose from. There are a few important things to remember when picking the right place.

First, you want to ensure that you are getting a dog from a place you support. Before adopting a dog from anywhere, whether it be a puppy store, a breeder or a shelter, do some background research as to how they train their dog. Of course, the dogs aren't at fault for how they are treated, but it is best to make sure that we are supportive of those who care for their animals properly.

It is also important to remember that puppies will always be in higher demand, but do not overlook older dogs! We will be focusing on training your puppy throughout the book, but don't forget that contrary to the popular saying, old dogs can learn new tricks too! Even if you adopt a dog at 10 years old that isn't potty-trained, has a problem with barking, and has almost no structure, you can still train them. The difference between an older dog and a puppy in training is that puppies are simply easier to train.

When deciding on *where* you will be getting your dog, keep in mind that every year, over two million dogs are given to shelters. Not even half of them will end up adopted (ASPCA, n.d.). There are plenty of puppies in shelters as well. Just because a dog isn't the size of a toy animal doesn't mean it is not a puppy! A puppy is any dog less than a year old. Between their first and second years is when they are considered adults, but even then they won't always act like an adult.

After about six months and up until eight months, dogs grow out of their 'puppy' looks as well. They get bigger, their ears and face grow into an adult's, and their coats are usually fully grown in by this point. Though puppies are adorable, we will only have them as puppies for six months, so it is essential that we make the right decision for our future, as well as our dog's. You have several options from where you can choose a dog. These are the most common places that people get dogs, according to AnimalSheltering.org:

- Adopted from Shelter (44%)
- Bought from Breeder (25%)
- Pet store (4%) (Animal Sheltering, n.d.)

The remaining were smaller and random statistics, such as adopting from a friend, received as a gift from a relative or taken in as a stray.

Let's start by understanding what a shelter is. The first kind are municipal shelters. These are run by government officials and the animals are usually pick up off the street. Compassionate employees ensure population of certain dogs and cats are under control. They also ensure that sick and injured animals are given the treatment needed. These shelters are often overrun, so as a method of controlling population, animals will be euthanized after certain periods of time, or under special conditions in these shelters. Adopting a dog from a municipal shelter might be the best method to save a life!

There are private shelters as well that will take in animals under some of the same conditions. They will also adopt animals from caretakers that can no longer handle them, and they might even partner with municipal facilities. Private shelters might euthanize, but many strive to be no-kill as well. It is up to you to discover private shelters in your area and do your research to see what their adoption policies might be.

Humane societies and SPCAs will also have animals up for adoption if you have one in your area. These are more commonly known for their advocacy for animal rights, promoting education or aggressive neutering and spaying efforts to keep population down. Some are no-kill and others aren't. It is important to remember that even in no-kill shelters, overcrowding can occur, meaning that sick animals might not get the treatment they deserve.

Breeders are usually smaller family homes that make it their job to breed the dogs they have to sell the puppies in a new litter. Breeders will usually have one specific breed, giving you the option to seek out a type of dog specifically. Breeders are much more regulated than they used to be, and many veterinarians will have recommendations for you to find one which is more reputable. If adopting from a breeder, make it your mission to meet them and the puppy before adopting.

Sometimes you might have to travel far to do so, but it is worth the trip. Sometimes, puppies are kept far from any humans and live in shelters or sheds away from the home of the breeder. This is a sign that the animal might not have had enough socialization. Of course, we want to help all dogs, especially those in bad conditions. However, if you have a family of five children who are still young, an anti-social dog isn't going to be good for you. The best way to ensure that your new furry companion becomes a lifelong part of your family is to make sure it fits in with you and your life in the first place.

A final option you have is through a pet store. These are very controversial because of the use of puppy mills in order to populate the store. The Humane Society of the United States conducted investigations under cover and made disturbing discoveries while using hidden cameras. They reported that there were "significant violations, including sick and injured dogs who had not been treated by a vet, underweight dogs, puppies with their feet falling through the wire floors, puppies with severe eye deformities, piles of feces and food contaminated by mold and insects (The Humane Society of the United States, n.d.)."

The first thought is that these puppies should be purchased to help keep them from having to live in these conditions! Unfortunately, what happens is that this increases demand, meaning that more puppies would have to go through this. Puppies that aren't purchased from the store will end up being given away if they haven't been purchased after a long time. You are entirely in charge of your own decision to purchase from a pet store if you'd like, but as a responsible pet owner, you should be doing proper research first.

This is a big decision for you to make, which affects both your personal and family life. There are many benefits and disadvantages of certain places, and the welfare of dogs is highly sensitive in many societies. For this reason, certain places might be labeled as immoral, but at the end of the day, it will be up to you to decide!

Energy Levels of Certain Dogs

Now that you have a good understanding of where to get your dog from, let's talk about factors that you will need to consider when picking your breed! The most important part for picking your breed is understanding their energy levels. This is because it will directly affect the kind of care they need. Some breeds might need a huge yard that they can run in on a daily basis. If they don't get this time outside, it could increase their aggression or make it difficult to control them when they are excited. Other animals might only need a couple short walks a day and are perfectly fine in social settings. To make sure that you are picking the right animal, you'll want to consider how energetic they are.

Individuals who work long hours with jobs that take up a majority of their day-time should have lower energy dogs. If you can't take your dog for walks during the day and only have time for a quick potty break at night, a large, high-energy dog shouldn't be your first pick. If you have children who can help care for the dogs and who will be playing with the dogs frequently, a high-energy dog is perfectly fine. Single individuals who live alone would probably want a low to moderate energy dog. If you don't have access to a yard and live in a smaller apartment, a low energy dog is first choice. Having access to a dog park or a regular park and beach is great for your dog and can help provide them with a lot of freedom and energy.

Consider their energy levels for when you are training them as well. Some dogs will be able to train easily and will know right away what you are commanding them to do. Other dogs might be harder to control because they are so playful and active. Being active or not isn't something that is good or bad. It will all simply depend on your own personal energy level and how much you'll be able to contribute to the dog training process.

Lower energy dogs are those that are considered toy breed dogs, and the non-sporting group of dogs. Toy dogs have the lowest energy. However, this doesn't mean they aren't necessarily active. For example, a Pomeranian has low energy because you won't need a large yard for it to run, but it could still be the loudest barker in your apartment complex! Toy dogs were bred to be a companion to humans. They are specifically used to be carried around and sit on the laps of their owners. They will be best for those who are looking for a dog to have as a friend. These are best for older individuals, those with mental illness that might want an emotional support animal, or single individuals that aren't as active. If you are single or live with a partner or perhaps you live in an apartment complex, this is the best category for you. It doesn't mean that your dog just lays around all day either! All puppies are very active and will require training to prevent barking, jumping, and other high-energy characteristics. However, as time goes on, low-energy dogs will be the kind that become more relaxed and require less physical activity than medium and high energy dogs. These are breeds typically known for having low energy:

- Bichon Frise
- Boston Terrier

- Bulldog
- Basset Hound
- Bloodhound
- Mastiff
- Manchester Terrier
- Pekingese
- Papillion
- Irish Wolfhound
- Chow Chow
- Affenpinscher

High energy dogs are those that stem from the terrier, sporting, working, and herding groups. These are dogs that were bred to be by your side at all times. Hunting dogs were trained to retrieve the flying animals (duck, etc.) that their owners shot down. Herding dogs were trained to help round up farm animals such as cattle or sheep. Herding dogs love having wide ranges to run around, and they will certainly try to naturally herd humans, especially children.

Book 3 - Training Your Puppy Step-By-Step

Working dogs are those that have been trained for protection. They might also pull things like sleds or be trained for rescues. Terriers were frequently trained in order to help protect the home and hunt smaller animals. Dogs from these categories will want to swim, play games, and spend a lot of time outdoors. They are very friendly and intelligent as well. If you are someone who loves spending time outdoors, has a huge yard, wants a dog for a specific task (hunting, fishing, farming, etc.) then you will want a high energy dog.

If you want a dog with high energy, these are dogs you should consider:

- Australian Shepherd
- Shetland Sheepdog
- Russell Terrier
- Siberian Husky
- Dalmatian
- Border Collie

- Labrador
- Golden Retriever
- Beagle
- Poodle
- German Shepherd

Medium energy dogs are those from all kinds of breeds but are known for having a nice balance between being able to play outside with families, but also be fine with sitting on the couch with you while you binge watch TV. This is perfect for those who have a fairly active lifestyle but aren't quite the kind of person who spends the majority of their time outside. These are also the dogs that are perfect for families who want to include a pet but might not have as much time for their animals, like a single individual or a young couple might have. For medium-energy dogs, you might consider:

- Australian Terrier
- Alaskan Malamute
- Akita
- Bernese Mountain Dog
- Bergamasco
- Bloodhound
- Corgi
- Chihuahua
- Cavalier King Charles Spaniel
- Cocker Spaniel
- Dachshund
- French Bulldog
- Maltese
- Many Terrier breeds

Consider all aspects of your life and the way that you do things now. Getting an active dog doesn't mean that you will become active yourself necessarily. Dogs can do a great job at teaching responsibility to different owners, but it is best you get a dog that fits in with your lifestyle rather than trying to fit in with the lifestyle of the dog you choose.

Child-Friendly or Not

Having a pet for your family is a great addition. It is another member who you can share memories and joy with. It can teach you and your children responsibility, compassion, and kindness. Some of the best memories that children have include their pet! When considering a pet for your family, there are a few things that you will have to remember.

Consider the age of your children. If you have teens, then you can assume they will interact with a dog similarly to adults. Children between 10-14 might not be as active, so they can still be fine with a family dog that works for you as parents. If your child is younger or petite in size, no matter what their age, consider this if you are adopting a larger dog, as they might still end up being on the herding end of things!

For any family that has a young and active child, especially more than one, then you have to be very careful when picking out a dog. Dogs are sweet, kind, and loving. However, they also have the ability to attack and kill. Dogs don't need to be feared, but we have to remember that just like humans, they can act aggressively when provoked. Although you could train your dog and your child to interact peacefully with each other, when your back is turned, you never know how a dog might react if the child tugs on its tail, flails its arms, screams or does something else that could trigger an otherwise peaceful animal.

Dogs will act on instinct, and just like children, they don't always understand the right social cues. At the same time, remember that this is a fact for ALL dogs, so getting a child-friendly dog doesn't mean that you don't have to have a discussion with your kids about the proper way to care for it. Even a child staring at a dog for too long can be enough to give it mixed signals! These are the types of dogs that don't work well in households with children:

- Shih Tzu
- Siberian Husky
- Greyhound
- Pekingese
- Jack Russel Terrier
- Chihuahua
- Australian Shepherd
- Rottweiler
- Saint Bernard
- Akita

These dogs are less likely to have issues when you adopt them as puppies and raise them with your family. Again, this isn't to scare you away either. Most of these dogs are fine, and with proper care and training provided to both humans and dogs, issues can be avoided. These are simply dogs that are more likely to play rough, nip and bite, and growl or scare dogs. Many herding dogs will nip at the feet or ankles of those running around, so it is not always that your child is at risk for getting seriously injured, but they might learn to be fearful of the dog. Pick one that is right for the whole family and give you and your kids the chance to meet your dog and play with it before adopting.

Interacting with Other Animals

If you are adopting a dog, consider your other animals as well. We will discuss breeds that work best with different dogs but remember that some shelters might not let you adopt if you already have a certain number of animals. If you do plan to grow your furry family, ensure that all the pets you already have are up to date on their shots and you have documentation because there's a good chance you will be asked to provide proof of their medical history.

Cats and dogs have long been believed to be enemies. Part of this is because they are such different animals. Cats are perceived to be devious, sneaky, quiet, relaxed, and reserved. Dogs have always been perceived to be more active, excited, and friendly. One isn't necessarily better than the other, though this is a hot debate. They are entirely different animals with various needs and qualities that keep them as separate species. At the same time, having one or more of each in your home is a great addition to any family. Though they can be enemies, they can also be best friends.

The reason that dogs and cats might fight is because of their natural instincts. Cats are naturally more afraid of things that are bigger than them. They also need personal space and time to get to know new animals. If you walk into a room and pick up the cat sleeping on the couch immediately, it will not be so happy about this. If you walk into a room and start petting a dog, they're more likely to start wagging their tails and feel excited about your presence. Cats struggle with dogs because they aren't given that space!

Dogs struggle with cats because they are smaller and quicker, much like the rodents that they have been trained to attack. A cat might move across your house the same way an opossum or a raccoon might. Though they aren't rodents, they are still animals that move around just like the ones that dogs have been bred to capture.

Typically, it will be easier for you to bring home a dog and introduce it to a cat rather than the other way around. The cat might be angry at first, but after some hissing and hiding, maybe a swipe or two at the dog, it'll warm up. If you bring home a two-pound kitten and introduce it to your 80-pound Tibetan Mastiff, you might struggle to train the dog to be dominant whereas a cat would rather hide than try to fight new visitors.

Cats don't respect territory as much as dogs do either. Most cats are independent animals whereas dogs are used to being in packs. Of course, animals like lions are frequently found in families, but your household cat is very different from a pride of lions than it is from some dogs! Cats will walk up to you and sit on your lap whenever they want. Dogs are more territorial and when they see this, they might be likely to try and protect you from the cat. At the same time, you can have both in your home! The best kinds of dogs to bring home if you already have a cat are toy dogs and non-sporting dogs. This is because they will be more likely to listen to you, their pack leader, and trust that this animal isn't a threat. Obedient dogs that are well-trained will typically do fine with other animals anyway, but still consider your cat before bringing home another pet. Listed below are the dogs that will do the best with other animals, and even cats:

- Maltese
- Boxer
- Bichon Frise
- Cocker Spaniel
- Basset Hound
- Labrador and Golden Retrievers

If you are considering having multiple dogs, you will want to consider different breeds that work well together, and those that don't. Many households will simply get dogs that are from the same breed, and this is the safe choice. We should still consider those breeds that don't quite mingle the best together.

First consider your dog's past and if they have gone through anything negative with another dog. For example, let's say that you already have a mini poodle at home, and you want to introduce another dog. There is a pug that lives in the house next door that you think is cute and you'd consider buying. However, the pug barks and growls at your mini poodle every time you go outside together. Though the pug might not have anything against either of you personally and is just being territorial, your mini poodle will still have a negative perspective about pugs. They will associate their ear shape, face shape, and body shape with something scary, so don't bring anything home that looks like a pug in this case.

Some dogs are more territorial and are the 'alpha' dogs. If you have a strong territorial male dog already, do not bring home another male dog. You would instead consider introducing a female since they are less likely to be aggressive. For multiple dog homes, you will want to consider males more often as singular pets or only have one male with other females, and have multiple dogs be female. This isn't the case every time. there are plenty of households with multiple male dogs that do just fine. In any case, if you are picking out new dogs to introduce to the ones you already have, don't overlook this point.

Small dogs will tend to be a little more aggressive toward big dogs. They will feel threatened and as such will show their dominance to the big dogs since they feel as though their size might be lacking. If you have a Pomeranian at home and want to bring home a Chow Chow, then you just might end up finding that the Pom doesn't stop barking and continues to try and exert its 'alpha' stance.

Consider dogs with matching needs over all else. You don't want to get a Pekingese, who could sleep up to 2o hours a day, and then bring home a Labrador retriever who will need hours of play. If your lifestyle is this diverse, it could work, but you want your pets to be on similar schedules, so it is best to consider that their needs will match each other.

If you only plan on having one pet, and one dog, that is perfectly fine! There are many dogs that just want to hang out with you and you only. The dogs that prefer to be alone the most are:

- Great Danes

- Collies and Sheepdogs
- Poodle
- Pug
- Chihuahua

Grooming Needs

Some dogs are fine with a quick rinse and a brush, but others need bi-weekly or monthly grooming. If you get a dog that needs frequent grooming, but you don't have any time to groom them, they can get knots, matted hair, and start getting stinky much faster. Of course, many individuals like the look of groomed dogs and wish to have one for aesthetic purposes. Grooming can be expensive and tedious, so you'll want to consider the different grooming needs of various breeds.

The dogs that need the most grooming can also be the ones that are best for certain families, however. Hypoallergenic dogs are those that don't shed as much, so they are best for people with allergies. At the same time, since they don't shed, their hair will get rather long. If you don't properly groom them, their fur can get knotted. Ungroomed hair can collect sweat, dirt, and bugs. This could harm your dog's fur and make it easier to spread pests inside your home. If a dog has a knot, they can still be bothered by it, so they might chew and scratch, causing a more serious injury than the initial knot.

On the other hand, if you have dogs that don't need regular grooming, at least from a professional, they will still need some form of care and maintenance. All dogs will still need an occasional bath, and it is important to keep up with their nail and tooth hygiene. We will cover that in a later chapter. Dogs that are non-hypoallergenic shed their fur rapidly because they are constantly growing new hair. This means you will have dog hair on your clothes, furniture, and everything in between.

These are the dogs which will require the most grooming:

Book 3 - Training Your Puppy Step-By-Step

- Poodles – all variations (mini poodles, cockapoos, etc.)
- Bichon Frise
- Shih Tzu
- Maltese
- Yorkshire Terrier
- Mini Schnauzer
- Lagotto Romagnolo
- Irish Water Spaniel
- Cairn Terrier
- West Highland Terrier
- Scottish Terrier

These are the dogs which won't require frequent haircuts, but will shed a lot:

- Corgi
- German Shepherd
- Australian Shepherd
- Labrador Retriever
- Siberian Husky

These are dog breeds which will frequently shed and also require grooming and haircuts. These are not recommended for those with allergies or those who don't want to keep up with grooming:

- Akita
- Chow Chow
- Collie
- Pomeranian
- Pekingese

If you want to get a dog that requires grooming, of course, you don't have to know how to do this yourself. It is quick and easy to drop your dog off at the groomer's or pet shops that provide grooming services. This can actually help socialize your dog when done frequently. The biggest downfall of this is that you need to find the right groomer, which can sometimes be expensive. The more work your dog requires, the pricier it will be. If you are already struggling financially, don't get a dog that will need regular haircuts.

Other Breed Specific Conditions to Remember

Many of us have an idea of a breed that we want years before we even get a dog! All dogs are special, and they each have their own personality traits you'll want to consider. There are a few other factors you have to remember when picking a dog. Some people will give dogs to shelters over simple things, so to ensure you keep your dog around for the long term, remember some of these important aspects when deciding a dog to pick:

- Breed-specific illnesses
- Dogs with separation anxiety
- Dogs more prone to health issues

Of course, we want our dogs to be healthy. Because of years of breeding, there are certain health conditions that some dogs have gained throughout the years. While it is incredible what we have been able to do for dogs including the way that breeding helps them have certain characteristics, it is also essential to consider how breeding has led to more serious health issues.

Larger dogs, such as German Shepherds, might be more prone to hip dysplasia. Boxers are also at risk for elbow dysplasia, among certain cancers and heart conditions. The same can be said for Labrador Retriever. Both Labs and Golden Retrievers are also more prone to be overweight, which could put them at an even higher risk for health-related issues. These dogs require a lot of exercise, but it is also important that we don't put too much pressure or strain on their bodies to keep their joints intact and healthy.

Book 3 - Training Your Puppy Step-By-Step

In addition to some of these health conditions, larger dogs are also known for not living as long as smaller breeds. Part of this might be because of the way that they have been bred. They grow so rapidly that it ages them faster in the process. The larger the dog breed, the less likely it is to have a longer lifespan. This is sad in itself but remember the impact it could have on your family. Some large dog breeds, such as Great Danes, might only live to be five to eight years old (Soniak, 2013).

Smaller dogs are known for living longer since they age much slower. They might live up to be 15 years old depending on the breed. However, this doesn't mean they aren't still prone to disease and illness. Some smaller dogs, like Yorkshire Terriers or Miniature Poodles are more prone to have issues with their gastrointestinal systems. Part of this might actually be because of the stress they experience! Some smaller dogs are more likely to have ulcerative colitis, which is when the intestine or colon, becomes inflamed. It is more treatable, but when untreated or unmanaged, it can cause more serious health issues.

Certain breeds, like Pugs and Bull Terriers, have been so bred throughout history that they are very likely to experience certain health issues. They might have trouble breathing, could become dehydrated or exhausted faster, and will even snore at night in some households (Kiefer, 2017)!

All these dogs are equally cute, special, and important. It is still crucial that we consider their medical needs, because vet bills can be massively expensive. If you can't afford to pay for certain surgeries upfront, you could stand to lose your pet. Some veterinarians will work with you on creating payment plans, but many will require that you have several thousand needed for surgery upfront. We should do everything to care for our pets, but if you are not in a position to pay $500 or more in vet bills unexpectedly, it is important to reconsider getting a pet.

Aside from issues with health, some pets will be a little needier than others and require more attention and care. Some have separation anxiety that could cause them to have deeper issues. No dogs enjoy being home alone, but they will get accustomed to it in due time and with proper training. However, sometimes, being left alone for hours at a time can cause them to have some serious anxiety which can mean they may have accidents indoors, chew at furniture or objects, and whine or scratch at the door for hours on end. Not only is this annoying for neighbors, but some dogs have actually scratched at doors to the point that their paws will bleed. Consider your lifestyle and ensure that you don't get a dog that doesn't do well snoozing on the couch alone for a few hours while you are at work. If you do get one of these dogs, make sure you have it in your budget to hire a dog walker or sitter to check in on the dog during the day while you are at work. Listed below are the neediest dog breeds that do not like to be alone:

- Cavalier King Charles Spaniel
- Australian Shepherd
- Labrador Retriever
- Jack Russell Terrier
- Border Collie
- Miniature Poodle
- Bichon Frise

Chapter 2: Preparing Your Home for Their Arrival

Your dog will no doubt be scared when you first bring them home. They're coming into a new environment they've never been in before, which can be rather frightening! Set them up for success by using care and consideration when setting up your home. Whether your dog is three weeks or three years when you introduce them to their new space, there are a few things to consider. Not only is the environment in which you'll be training them important, but the tools that you have are special to consider as well. Stay positive, create healthy habits, and hide anything you don't want in the mouth of your pup! Let's dive deeper into these topics so you can be as prepared as possible.

Starting Positive Habits from the Beginning

From the moment your dog walks into their new home for the first time, you should do your best to train them to have good habits. If you don't start right away, then the dog won't be able to pick up on these healthy habits you are trying to instill so you can elicit good behavior. Dogs are very ritualistic animals and prefer to have a routine. Not only is it their way of operating within our society, but it also gives them peace of mind. Dogs can be very anxious creatures, especially if they can't predict what will happen next. A routine gives them the chance to understand what they might need to be prepared for and the ways that they can best go along with the natural flow of things.

Don't put off any vet appointments and necessary shots. Your puppy might seem perfectly healthy with no issues whatsoever. You might not be an active person or someone who would take your dog out in public anyway. In this scenario, it can be easy to think, "Nothing bad is going to happen, the puppy will be fine." Chances are, this is true. However, after all the love you give to your dog, this should be an extra step to ensure that they will be around for a long time. You never know what other dog they might interact with who could spread something. They might get one flea bite which could have been from an infected insect. There's no point in risking your dog's health, so keep up with necessary vaccines from the moment you bring them home.

Make sure to purchase all that is necessary before getting started. The dog should be the last addition that you have for your home and family. They should have all the necessary toys, bowls, leashes, collars, harnesses, beds, and everything in between to ensure that they will feel comfortable in their new home.

Practice what it will be like to have a dog before you even bring your four-legged friend home. It can seem tedious, but it is a good way to ensure that you are really prepared for the routine. Start by making sure that you keep all your belongings picked up. Puppies are like toddlers and can be great at finding random objects to chew on or play with. Every three to four hours, make it a point to go outside for a quick walk. Make sure that you get in the habit of establishing set times and routines so you can leave the house for about 10-15 minutes every three to four hours, just as you would once you start walking your dog on a regular basis.

Check-in with your emotions, because dogs require a lot more than just physical energy. Make sure that you are a patient and compassionate person. If you are someone who struggles to keep their cool, especially in stressful situations, it will be good to come up with some coping mechanisms because puppies will certainly test your patience. There will be moments where they have accidents, they won't always fully understand what you are trying to say to them, and they'll cause frustration among some family members. This isn't a bad thing, but it can be challenging to deal with, especially for individuals who don't have their emotions in check.

Be prepared to put in the work all day long. Even though your puppy will have plenty of moments of playing, sleeping, and just relaxing, you will still have to be alert. If you can't see your puppy, you'll have to find them to ensure they didn't get lost under a piece of furniture or in a different room. You'll have to be prepared to wake up early to take them out, fill their water bowl, and feed them. If any of this sounds challenging, then there needs to be more preparation done on the owner's end before bringing a new puppy into the home.

Make sure that you have the proper discussions with roommates and other household members. Decide who will be in charge. For example, if you have a roommate who you aren't very close with and who isn't really interested in dogs in general, you can't expect them to be a caretaker. If you are bringing a dog home to a family with a husband, wife, and three kids, then you'll probably want to split the responsibility equally among everyone. It is important to have a talk about all responsibilities beforehand. You can create a schedule for family members. Maybe each day, a new person is responsible to take the dog outside. Perhaps one person is in charge of feeding the dog and everyone else has to take turns walking the dog. You have to have these discussions before, so you are prepared instead of being stressed and confused in the middle of it all.

There has to be consistency when training the dog. That is key. You'll have to make sure that there are house rules. Mom cannot let the dog on the couch if dad wants it to always sit on the floor. The kids cannot give the dog any snacks that they're eating if the rule is that the dog only eats dog food. One family member cannot have a rule that the rest of the family members don't follow because this will hurt the dog in the end.

Traveling isn't going to be easy anymore. Make sure that you don't have trips coming up, and if you do, ensure that there's someone there who can take care of the pup while you are gone. Make sure that you have a schedule so that the puppy is never alone for more than a maximum of three hours. For puppies younger than 12 weeks, it should really be no more than an hour. Once they are older than a year, and in some cases six months, they can stay home for longer and up to six to eight hours depending on the dog breed.

There will be a lot of changes. Nothing will ever be the same. Some things you'll be prepared with, others will certainly surprise you. Don't underestimate just how prepared you have to be!

The Right House Setting

Your home is your puppy's home. It has to be a welcome and friendly environment in which they can thrive in. Imagine that you are the puppy before bringing it into your house. You are currently with all your brothers and sisters somewhere trying to figure out how to walk, run, and even bark. Then, some new people you never met pick you up and carry you away. They bring you to a place that you've never been to before. What would you want in this new place to make sure that you feel right at home? We can't project emotions on puppies like humans, and animals don't stick with their families all their lives anyway. What's most important to remember is that they will be frightened because things are suddenly different from how they just were. Ensure that your puppy is given all it needs to feel right at home from the moment you bring them home.

The puppy needs its own space. As it grows, you won't have to dedicate an entire room to the dog. You might be able to eventually just have them roam the house, even when you are not there to watch over them. As a puppy, you should set aside a dedicated area where they can run freely, sleep, and nap during the day as you are with them. Don't give them free range of the house just yet. That will be too overwhelming, and they'll end up getting themselves into a place that you didn't even think of at first. Your puppy will want something familiar, so use baby gates to block off a section of the house that they can feel the safest in.

Give your puppy a place to play that doesn't have a rug that you don't want ruined. You can't bring a puppy home and into your freshly carpeted living room, only to get mad that it immediately pees on the carpet. If you do have a carpeted room, invest in some thrift store sheets or extra rugs to put over the floor to protect it from the puppy. At first, your puppy won't defecate or urinate in massive quantities that are hard to clean. It is still important to be prepared for them to go 'potty' wherever and whenever they have to as you are starting to potty train them.

You'll want something steady and constant for it to bond with. Create a little room in an office, a bathroom or laundry room for your pup. These are smaller areas that they can go in for naps and get away from it all. Block off stairs just like you would for a newborn baby. Anywhere that a small creature could get hurt is an area that needs to be entirely inaccessible to your puppy.

Use puppy pads to create a path to the back door or the door that they'll be regularly going out to go to the bathroom. Let's say you are creating a space for your puppy in the sunroom that is adjacent to your living room since there is a direct path toward the backdoor on the other end of the house from your living room. Additionally, you spend most of your time hanging out in the living room, which will make your puppy feel less lonely when they spend time in the sunroom. You put a little bed in there, their crate, and some water and toys. Use newspaper or purchased puppy pads to create a path all the way from the sunroom to the backdoor. Each time they have to go out, let them walk along this path so they get a feel for where they're supposed to be going to the bathroom.

Make sure other animals have their own spaces during this time as well. Your cats will especially be more scared, so you will want to put their litter box or food bowls in a private space so that they can have alone time as well. Don't do anything that will freak them out, but if their litter box is in the laundry room where you also plan to keep the puppy, then they might not want to go in there, and could end up peeing or pooping in a random spot in your home. Consider all the animals and their need for their own room, not just your puppy.

Tools Needed

Once your home and your family are prepared to welcome your new furry companion, it is time to start stocking up on the right supplies. The best place to purchase supplies are from a certified pet store. You can shop online as well. When you go to certain big stores that stock cheap and generic dog supplies, you might discover that the products aren't as high-quality. This is fine if that's in your budget, but there should certainly be an emphasis on purchasing the highest quality items for your dogs. It may seem expensive initially, but if you really think about it, you will only need to buy most of these supplies once. For example, you'll purchase food and water bowls for them, but it is not like you'll be buying a new bowl every month. Every year or so, it is good to replace things just because you never know what bacteria or other germs they're holding onto that don't get washed enough, no matter how much you clean and disinfect them. However, most of these things will be a one- or two-time deal to buy, so now is the time to splurge and spend your money on your four-legged friend!

The first thing you'll want is a collar, harness, and leash. At first, puppies won't be able to run far away from you, but this doesn't mean you shouldn't still train them on a leash early. Make sure that you buy a collar specific to their size. You should be able to fit two fingers sideways into the collar to make sure they have enough room to breathe and grow. Get them the appropriate tags as well. Their name, along with a phone number and even your name is a good choice so that others can call them and call you if the puppy ever gets separated and lost. Tags with their vaccinations will also be helpful so that others know if the puppy is at risk of carrying or catching any diseases or illnesses.

A harness is a good choice as well. Puppies especially will tug and pull on their leash, and only having a collar might mean that they hurt their throats frequently. As you start to go for walks, the puppy might frequently get into things they shouldn't. Maybe there's an old sandwich or chicken bone that was tossed on the street from someone who was walking by earlier. This is a big find for your puppy and they'll try to devour the snack! Rather than jerking their necks back on a collar to stop them, you'll have a harness that could even help you to pick them up quickly.

Training pads will be a good addition as well. These are items that don't necessarily need to be the most expensive brand, and many stores carry generic versions that do just as well. The point is that they absorb moisture and provide your puppy with a scent to let them know that this is the place they should be using the bathroom, nowhere else. These are not to be used in replacement of going outside. If you want to train your puppy to use pads even into adulthood as they become full-grown dogs, that's perfectly fine. Some people might live on the 20th floor of a building and going outside three to four times a day isn't an option, especially late at night in the big city. They might want to use pads, and that's fine, it is totally up to the individual. However, if you do plan on training your dog to only go outside, the pads should be for accidents, not the only way that you teach them to go to the bathroom. They should be used during a transition period as the dog becomes accustomed to going outside.

You need to also purchase dog food and treats, but we'll get into the specifics of this later. As far as food bowls go, stainless steel is always the best option. Plastic bowls can trap moisture and cause bacteria to grow, potentially making your dog sick or even giving them little pimples on their chin! If you are using a glass or porcelain bowl, it could end up breaking if your puppy gets too rough. Stainless steel is best, especially since it can be cleaned very easily. Purchase one that has a rubber bottom to keep it from sliding all over the place as your dog eats. Keep it low at first for the puppy, but depending on the breed, a raised food bowl might be better so that they don't have to strain their neck or back to reach down to eat.

Have some form of a pet bed for them. It doesn't have to be a big dog bed, it could simply be a cardboard box with some blankets and a pillow. You could give them a few different options as well to see which bed they like the most. They simply need a place where they can sleep all alone and do so whenever they feel like it.

Toys are a big requirement! You want to teach them to play with certain objects and not to play with things that are more important to you. We'll help you discover the way that you can find what's best for your furry friend. Let's discuss now the option of using crates as a training room for your dog.

The Training Room for Your Dog

You have two options for your dog's 'room.' You can use a crate or you could have an entire room dedicated to them. A crate is preferable to most families, but if you can dedicate an entire room such as a spare bedroom, this is always ideal. Don't use something like a mudroom where people will be passing through frequently. For this section, we are mostly going to be discussing a crate as your dog's training room.

The reason that your dog needs a room is because they are den animals. This means that they are used to finding small holes and caves that they can associate with being their own. Think of animals in the wild that are like dogs. Wolves frequently find caves which they can easily hide in. They might find an old animal nest and use this as well! Your dog's room can be like their crate. If you choose to use a crate, then you will want something much bigger than the dog. For example, look at the photo above. It looks like it could probably fit into an even smaller crate, but it shouldn't. The dog should be able to stand up and fully turn around in their crate. If they can't, it is beneficial to purchase something a little bigger for them.

This room is not something that you should use to simply shut them away in. It has a specific purpose. This is the room your puppy will use to relax, play, eat, and sleep in, along with their puppy training. It is not a form of punishment and it should never be something that you use to isolate them. It is a safe haven, a place they want to go, and somewhere they feel completely comfortable in.

This is the room that they will go to when they need to sleep or when you aren't going to be home. A place of comfort and familiarity. If they have been playing for over an hour and you trained them throughout, you can put them in the crate as a way to let them know that this is a quiet time.

Never use their crate as punishment. If they go inside the house, chew on something they're not supposed to or engage in any other negative behavior you want to punish, don't put them in their crate. They won't be able to understand that what they did is wrong so therefore they need time to reflect and think about things. They will simply see that you are upset and confused about the situation.

Don't put your dog away for more than three hours at a time when they are younger than six months old. As your dog becomes an adult with more bladder control, then you can crate for up to eight hours, but this should really be the maximum. This should only be in rare circumstances when you can't find someone to help them out, not the entire day. If you are someone will be gone for eight hours and beyond, using a crate shouldn't be your puppy's room. You can use the crate and give them the option to go in, but a bigger space is better if you ever have to leave them for that long at a time.

Your dog wouldn't go to the bathroom in their den, so don't expect that they'll just go if they really have to. Instead, they will hold their bowel movement to the point that they physically can't anymore. The den isn't a place to put them to go to the bathroom. It is an area where you can train them to sleep, eat, and relax. If you want to leave them home all day and don't care if they roam freely in your home, then using an entire room with puppy pads is a better option to keep them away.

This is not prison. Only keep your dog in its crate as you would want to go in your room. If you make them fearful of the crate even once, they will be scared of the crate every time that they go in it afterward. This is a place where they can be alone, have their private time, and stay completely relaxed.

Even if you want to sleep with your puppy, they should go through a crate training period. This is to help them control their bladder. At night, you can put them in the crate and be sure that they will try to hold it in all night. If you let them sleep with you, they might jump off the bed and wander to a spot to relieve themselves, then come back to bed without you ever knowing (until you find the mess, of course)! After a few weeks of crate training at night, you can start letting them out of the crate and see if they are able to sleep with you while holding in their bladder movements all night long.

At first, this room should be somewhere they can easily access where you and the rest of your family will be. Don't put their crate in the basement where they can't get to it. Keep their crate in the living room where everyone hangs out or the kitchen where you might always be cooking. This is because you don't want to create the idea that their crate means isolation or that they can't be around others. If you get them used to them being in their crate while you are still there, it will help them better understand that this is their home, not their punishment. Now that you understand the importance and meaning of this small space, let's discuss how to introduce them to the crate. Again, if you don't want to use a crate, a small room is just as fine. We'll be discussing how to use the crate and introduce this idea to your puppy, but by all means, apply this to a smaller room as well.

You'll start by introducing this crate to them just as you do everything else – slowly and led by them. Put some toys in there and let them go in on their own. Maybe you place a towel in the crate that has the texture of their mother to make them feel more comfortable. Some people have even tried to put clocks in there to mimic a heartbeat. You might even put a treat or two in the crate. Let them go in on their own and explore this place for themselves. It will be a place they can get to know and explore, not just an area where they are shut away.

Never force your dog in the crate if they are actively resisting. If it seems like it is something scary, be patient. Sit next to the crate and play with toys. If they aren't into the crate, don't pick them up and put them in. If they go in just with one paw at a time, don't push their butts trying to get them to go faster. It might take a few days for them to actually go in, but that's totally normal. It will be better for you and your puppy if you are patient in this process. If they don't go in, place a toy at the front of the gate, not all the way at the back. Let them get this toy out if they want, and even if they don't go in the crate at all, it is still helpful to make sure that they are slowly transitioning into this.

You can make this transition easier by giving them meals in their crate. Start by introducing them to the crate on the first day. On the second or third day, once they display more curiosity, put their food bowl right inside the crate, just at the front of the door so they could even eat their food without going all the way in. Each time you give them a meal, push it a little further from the door so they have to go deeper into the crate. Eventually, you can put the food all the way to the back of the crate.

This is when you can start to close the door while they are in the crate. Don't slam the door and startle them. Slowly close the door behind them and see how they react. If they panic, open the door immediately and use positive reinforcement once they come out. If they don't even notice,

Book 3 - Training Your Puppy Step-By-Step

congratulations!

Keep your dog in for 10 minute intervals, but nothing longer than that. After they've finished eating, let them stay in there for a few minutes, then let them out. Do this during every meal time until they don't show any signs of being afraid of their crate. Start leading them in without meal times after this. For the first week, do so every day. Practice letting your dog in the crate and keeping them in there for at least five minutes, but not much more than 10 minutes. This is to get them used to it. You should always be there in the same room during this training period. You don't have to be engaging with your dog. In fact, you should do the opposite. Watch TV, read something or just relax on the couch by them. Let them see you, but don't feel the need to look at them and play with them while they're in the crate. After these few minutes, let them out and use encouraging reinforcement the whole time, so they know that this crate is a positive experience.

Don't let your dog out because they are whining. Wait until a few minutes have passed after they've stopped whining to let them out. If you don't wait, they will associate whining as their key to getting out of the crate! It is hard to watch your puppy cry, but it is okay! It is better for them in the long run.

Practice this crating method before you actually leave them for long periods. Until they're used to a crate, you'll want to be home with them. This is why you should take a few days off work to make sure you are home with the new puppy.

You should be crating them when you leave and when it is nighttime, but not only these times. Crate them when you need to clean the house, or if you need to get some work done. Shorter periods are better, but if you only put them in the crate when they're alone, then it will make them associate the crate with isolation (The Humane Society, n.d.).

Hiding Certain Belongings

Puppies are like babies. They will explore and try to put things in their mouth as often as possible. It doesn't matter if it is a $300 pair of shoes or some poop from the litter box. If it has a scent, the puppy will try to see if this is food! You have to watch your puppy like you would babysit a child roaming around your house. A puppy has heightened senses, so they'll be even better at picking things out that shouldn't be eaten!

Make sure all electronics are picked up. This includes wires connected to the TV. Any cords or cables freely around the house should be put up. Use plastic hooks that can easily be removed from your walls to help keep wires and cords off the ground. Not only is this so they don't chew on them, but you never know what area of the floor might have a scent that triggers them to relieve themselves right there. If you can't put the cords up for whatever reason, you can use tape to keep them stuck to the floor and protected from your pup.

Block off electrical outlets. If you can't use a specific plastic child/pet safe outlet cover to protect them, use a baby gate to create a barrier to heavy electronic areas in your home.

Ensure that anything you don't want in your dog's mouth is put away. This is anything from shoes to throw pillows and chair legs, and so on. If you don't mind a little puppy slobber on your furniture, then keep it around, but anything else, make sure to hide from the pup. This won't be the case the whole time. Once your dog learns some rules and boundaries it'll be easier to ensure that they aren't getting making a mess or chewing on what they shouldn't be.

Check for loose buttons on furniture, thread getting pulled out, and other things that might be easy for your dog to put in its mouth.

Consider the cleaning products you might have used on the surface of certain things. Did you just bleach the bathroom? Did you use wood polish on the furniture? Your puppy might try to chew on these surfaces, and you don't want them accidentally licking up any harmful chemicals.

Plants can be an area of interest for pets. Ensure that any plants within the reach of your puppy have been put away, or that they are at least safe for animals to chew on and not something specifically poisonous to dogs.

Keep other toys that are meant for bigger dogs away from your puppies. This might be bones, big rubber items, and anything else that can be hard on their teeth or cause them to play too rough. Puppies need puppy toys! Consider things that you would never assume your dog has interest in that they still might. You might think, "That's fine sitting there, why would my puppy have any interest in that?" only to discover that it is in their mouth five minutes later.

Of course, there are many toxic foods that your dog should stay away from. These include:

- Chocolate
- Caffeine
- Grapes
- Onions and garlic
- Alcohol
- Raw meat
- Meat bones

Dogs love to chew on bones, but these should be safe bones purchased specifically for dogs. Chicken bones are more brittle, especially once cooked, so they could snap in half and your dog might end up choking on it. Even if you think you have managed to pick everything up you need to, you might discover that the puppy helps you see things that weren't there at first.

Choosing the Right Toys

All dogs deserve to have plenty of fun toys to play with! Toys help them learn how to interact with others while also providing them some entertainment. Dogs need to feel like they have a purpose. They like tasks because it makes them valuable members of their pack. They prefer to be outside chasing mice, birds, and other little critters to bring to their pack, but that's not ideal for us as human owners! Provide your puppy with all the toys they need for the best development possible. The best option for purchasing toys is to do so at a pet store. Be mindful of the quality of toys you purchase such as durability and materials.. Cheaper toys might have plastic or fragile rubber that breaks apart after just a few minutes of chewing.

Don't assume kids toys are safe for dogs. Just because a baby can play with it doesn't mean that it will be safe for your dog too!

Use toys that are specifically toys. Don't give them an old pair of shoes just because you don't care about the shoes anymore. They will think that all shoes are things that they can play with, not just the specific pair that you gave them.

Consider toys based on what your dog's breed is. Some dogs are hunting dogs that care about catching game. Toys that mimic animals like birds and ducks could be good for them. Other dogs are more interested in running and chasing, so they might do better with balls and toys that you throw long distances. You can never go wrong with a game of fetch! Many dogs like the same kinds of things, but it is also important to check with your specific breed and understand what their favorite things to do are.

Puppies want soft things they can roll around with. Their teeth are still developing, and they'll also want things that mimic their brothers and sisters. Puppies play with their siblings to learn how to function as adult dogs. They need that interaction to help them learn and grow.

Before they're six months, they will be teething pretty hard. Rubber toys and other things they can chew on will be best for them. Older dogs should have softer toys as well, as their teeth will be more sensitive.

For large dogs, don't give them anything may break right away. Even though it might be a specific plush toy made for dogs, your pup might have tough enough teeth to rip through anything they want.

Give them a variety of toys, not just one. This will help you discover what they like and dislike as well.

Dogs can have puzzle games too! You might get them a ball that has treats in it, forcing them to move the ball around to have treats fall out.

It is important to have toys that they can play with, however you want to ensure they socialize with kids as well and can play with them too. Some ways to ensure this are by including ropes that you could play tug-o-war with and having your kids throw some balls that they can fetch.

Now that you have all the supplies you'll need to help your dog grow and develop, let's look at the actual training practices you will be teaching your dog.

Chapter 3: Principles to Train Dogs

You have your puppy, you have all purchased all your supplies to ensure your puppy is happy and comfortable, and have gone through any other measures to ensure they are healthy and safe. You are prepared. Let the training begin! There are a few essential principles that will be important to teach your dog. You should follow these rules and do these things as best as possible to give your dog a chance to be as happy and healthy as can be. Potty training will be the first thing you teach them. This is because it is a basic need that they have, and there's no delaying their natural urges. In this chapter, aside from potty training, we will give you a few basic training rules as well as methods of positive reinforcement to use throughout the training process.

Potty Training

This is the most essential thing that your dog will learn. We will spend a lot of time in this section, because this is one of the biggest headaches for new dog parents. A dog that chews on some furniture isn't a big deal, especially because puppies haven't developed strong teeth yet. Having to teach your puppy some boundaries can be simple as well. The biggest headache is when a dog goes to the bathroom inside the house. It is stinky, inconvenient, and generally annoying. Don't worry or fear this though! It is still natural, and plenty of dogs are easily trained to the point that they don't urinate or defecate inside at all and haven't for years at a time.

There are a few important things NOT to do to make this transition much easier you are your pup. First and foremost, never push their face in pee. This idea had been around for a long time as a way to train your dog. People thought that if you made your dog uncomfortable in this way they wouldn't go to the bathroom inside the house again. This is completely false and could even delay proper training. You are simply teaching your dog that you are dominant and that they should be ashamed for going to the bathroom.

Making sure that your dog is properly trained isn't just about taking them out quickly when they have to go. To help prepare them and set them up for success with potty training, ensure that they are on a consistent schedule and your communication is consistent as well. Dogs need regulation. They don't know the difference between inside and outside. They know the difference between going to the bathroom in an area that's acceptable and somewhere they shouldn't go. The only way that they'll know what's right and wrong is by you teaching them the boundaries they have to follow.

You'll want to ensure that you help them have a schedule by feeding them at the same time every day. If you are feeding them at the same time every day, this also means they'll likely be going to the bathroom at the same time every day. It will be easier to predict their schedule so that you know when they need to be taken out. The dog will appreciate this as well because it will make it easier for them to hold their bladders when they have to.

Having a crate is the best option in the process. They don't want to soil in their cage. They like to keep their environment clean just as humans do. Cleanliness isn't about appearance. Most animals are aware that it is simply not good for your health to sit in urine or feces. Never let them go in the crate or expect that they will go in their crate just because you leave them in there for a long time.

Your puppy will be able to hold it for as many hours as they are in months, until they reach nine months old. This means that:

- A two-month old puppy can hold it for two hours.
- A three-month-old puppy can hold it for three hours.
- A four-month-old puppy can hold it for four hours.
- A five-month-old puppy can hold it for five hours.
- A six-month-old puppy can hold it for six hours.
- A seven-month-old puppy can hold it for seven hours.
- An eight-month-old puppy can hold it for eight hours.

Eight hours is the maximum that you would want your puppy to be holding it for. Anything longer could hurt them. They might be able to hold it for longer, but that doesn't mean that it is healthy for their body overall. Some puppies do take longer, so don't be upset if your six-month old puppy can only hold it for four hours or less.

You will want to feed them in smaller bits throughout the day rather than one big meal at once. Puppies should be eating around four times a day, and as they grow, you can cut it down until they are eating two meals a day. We will discuss food and feeding in a later chapter. It is simply important that you are making a list of everything your dog does. Even if it is something as simple as:

- 8 AM: peed outside
- 8:30 AM: ate
- 10 AM: peed inside
- 10:20 AM: peed and pooped outside
- 11 AM: ate
- 12:50 PM: peed outside
- 2 PM: pooped outside
- 3 PM: ate
- 4 PM: peed and pooped outside
- 6 PM: peed inside
- 8 PM: ate
- 10 PM: peed and pooped outside

As you create this schedule, you will notice a pattern. Then, it will be easier for you to know when it is best to crate and when you shouldn't, for example, you might crate them around 9:50 or 10 PM for a little bit as it seemed like they went to the bathroom within a shorter time interval. This is an instance where they need to be bladder trained, as they probably could have held the pee that they released inside at 10 PMuntil 10:20 when you took them out.

The best method of potty training is to simply take them outside around every two hours. Take them outside before feeding because they might not want to eat if they have to use the bathroom really badly. When they do go outside, use positive reinforcement. Take them outside as often as you can, so if you can do so every hour, that is even better. Give them a treat, and verbal praise along with a pat on their head or rub their belly so they associate going to the bathroom outside with good things.

If your dog does soil the house, do not yell at them. First, try to distract them. Maybe you call their name, clap loudly or squeak a toy. Distract them as you notice them about to go, and then take them outside as quickly as possible. If they do end up soiling your home and you discover it moments later, simply clean it up as best as you can and take them outside. If you yell at them as they are going to the bathroom, even if it is indoors, they will associate going to the bathroom as being bad, not just going to the bathroom inside as something that's bad for them.

When you see the behavior, try to make sure that you stop it. That is the only thing you can really do. This is why some people get frustrated. They want a way to punish the dog immediately in order to prevent it from happening again. The only thing you can really do is make sure you are taking them outside as often as possible and stopping them with distractions when they do want to go inside.

Clean it as best as you can to get rid of the scent. If they urinate on a certain part of the carpet, you have to do everything in your power to get this scent out or else they might go right back to it to pee once again. If any other animals have ever urinated or defecated in the house, then you'll need to make sure that this is cleaned up as best as possible as well. Dogs have amazing noses that can pick up on a lot of various scents. If you aren't careful with clean up, they will be able to pick up the scent and think that this is the spot they're supposed to go to the bathroom. There are plenty of specific cleaners that you can use which are made for the exact reason of covering up another dog's scent.

Take them on regular walks. When they go outside, give them praise. This is when you could use a clicker for training. A clicker is a handheld device that makes a clicking noise. Give them a treat and then press the clicker. When they go outside and go to the bathroom, use the clicker and give them a treat. It is a way to keep them focused on the good things that they do. They will know that the clicker means they did something good. When they go outside and hear the clicker, this means that they deserve praise. This is a method that works for many, while for some, it doesn't work at all. Nonetheless, it is a way for you to try to better teach your dog all about positive association.

After you've been taking them outside for a while, examine their stool and urine as well. Is their urine in small amounts and bright yellow? They could be dehydrated. Is it a ton of water and extremely clear? They might be drinking too much water too fast rather than sips spread throughout the day. Is their stool runny and hard to pick up? Is it hard and giving them trouble as they try to poop? These are all going to be signs that they need different foods or treats. It can seem tedious, but it will be really helpful to have a strict schedule you use to keep track of when they're going and what their stool looks like (Bovsun, 2019).

Basic Training Rules

There are a few commandments of training that are important to remember. We will discuss the important things that you need to always keep in mind not just for training, but about dogs in general. They are a different animal, and that's what many people fail to realize. We understand that they can't do certain things like drive cars, walk on two legs, and get jobs. However, too many people put human emotions on dogs. We think that they have the ability to feel guilt, reflect, learn the boundaries that humans abide by, and so on. While they can learn rules, they won't be able to emotionally grow like humans. When you make a mistake, you think to yourself, "Ah, I shouldn't have done that." You might blame other people or you might take responsibility and change for the better. When you go through that situation again, you'll know what to do to not make the same mistake. A dog doesn't have this thought process. They're mostly thinking that one thing equals bad and another thing equals good. They will simply focus on doing things that are good, things that please you because you are the leader of the pack. All they want is to earn your respect. We can't read a dog's mind, so who knows, maybe they are better at existentially reflecting and growing. However, this won't help us in training them. We have to be patient and remember that they simply don't think the same way that humans do.

Remember that your dog does not have the same sense of time that we do. They can't look at a clock and see how long it has been. Even humans don't always have the same sense of time in all situations. Think of how the ten minutes of extra sleep you get in the morning can go by in the blink of an eye but waiting in line at the DMV for 10 minutes can feel like three hours. Unless we actively monitor the clock, we don't always have a perfect sense of time. The same can be said about your dog. They don't even have the ability to look at a clock and measure time numerically. If you go to the store for an hour and a half, this can feel like the same time that it takes you to go for a ten-minute walk on your own. Dogs won't know the difference, so we have to remember that in training.

Dogs are not going to fight you back unless they feel as though their life is in danger. They can be naturally aggressive, but not because they are evil, vicious animals. They have a combination of instinct and feeling that will feed into their aggression levels. Dogs are only aggressive because they are scared. They need to learn how to play as well. If a dog is playing rough, they are simply learning the boundaries between play and real kill or attack. If they are aggressive and growing or snarling, they are likely just scared and trying to protect themselves.

If you are playing with your dog and they get to rough, make a loud and clear noise so they know that they are hurting you. Don't scream and yell, just a quick and sharp, "Stop" with an assertive over emotional tone is all that's needed. Some people will even yelp in the same way that a puppy would in an attempt to teach them about boundaries on their own level. The noise that you choose to make is up to you. Remember, you don't need to punish them. Simply stop the aggressive fight by letting them know they hurt you.

These dogs look up to you. You are their hero, their leader, the person in charge. Don't abuse this power that they've given you. Use it for good and train them to be happy and healthy dogs.

They will pay the most attention to your tone. Though they can't know exactly what you are saying, they will be able to pick up on your tone of voice to get a sense of whether or not you are happy or if you might be stressed or scared.

If you are stressed, your dog will be stressed. In fact, one study discovered that some dogs had the same cortisol levels that their human companions had. This means that they will pick up on the emotions we feel and then end up feeling it themselves.

Dogs are incredibly energetic, but they won't always be willing to run around and be animated. Remember that they need time to rest. Just because they don't want to get up to go to the bathroom doesn't mean that they're disobedient! We all know that feeling of laying down and not wanting to get up even if we do have to pee really badly. Whether your dog is a puppy or an old dog, they have the same kind of rules that apply to how you should be treating them.

Book 3 - Training Your Puppy Step-By-Step

Positive Reinforcement and Remaining Patient

You will want to become the pack leader for your dog, and any other pets that you might have. You can earn their trust through ways other than violence. Many individuals believe that physical punishment is the best way to teach dogs the rules. This is completely false, and studies have been proving that physical punishment over positive reinforcement actually hurts dogs in the end.

There is one important rule that you will have to live by among everything else:

<u>Never hit your dog</u>. This can't be reiterated enough. There are many people out there who might tell you otherwise. They might say, "It is their language," "It is how you become dominant," "It is how you show them who's boss," or "It is the only way they understand." Maybe it is a way to communicate. None of that matters, however. What we do know for certain is that dogs do not react well when they are hit. We know that it causes long-term damage, it physically hurts them, and they don't understand you. There are a million other ways to communicate with your dog. Screaming, hitting, kicking, and pushing your dog are not the ways that you are going to want to train them.

Know the difference between assertive and aggressive. When you hit your dog, it usually comes from an emotional place. Screaming might indicate frustration. There are emotions attached to these actions which dogs do not always understand. Separate your anger or frustration from the dog and do not take it out on them. Maybe they did poop and pee all over the house, got into the trash can, and chewed up your favorite pair of shoes. This is all terrible, but hitting them doesn't fix that, and it doesn't make them learn anything valuable. They will be scared as it is when they see that you are upset, and being aggressive toward only makes it worse.

Studies have proven that using punishment against dogs will increase anxiety and fear within them (Becker, 2012). They don't know that you are upset at them. They don't understand that you are trying to teach them a lesson. They will be confused and sad that the person they love more than anything is angry with them. Using aggression on them will make them aggressive in response to the same kind of stressful stimuli.

In fact, there is no increase in obedience in dogs who have been trained physical punishment as reinforcement for negative behavior Instead, some dogs are less obedient after experiencing repeated physical abuse. Use positivity for your training procedures. The way that they are going to learn better than any other dog is when you choose to use positive reinforcement with them. This means that rather than punishing them for the bad things they did, you'll reward them for the good things. If they are being bad, you can distract them and change their behavior by using a toy or another noise to pull their attention. If you don't catch a dog in the act of doing something disobedient, you have to wait until the next time to teach them this is wrong. If you get home from work and discover that your dog went through the garbage, they won't know that a punishment at 8:00 pm is for something they did at 6:00 pm. They have moved past this and forgotten.

Book 3 - Training Your Puppy Step-By-Step

Your dog should feel comfortable and trust you, not fear you. Put yourself in their paws, so to speak. Who do you feel comfortable around? Who do you love and trust the most? Who inspires you and encourages you? Who accepts you and loves you just the way you are, flaws and all?Then think of someone you are scared of, someone you cannot be yourself around, someone who makes you feel uneasy. Who is more effective in inspiring you? Which person do you appreciate more? Sure, your dog might listen to you if they are scared of you, but the bond you create with them will be built on fear, not love and trust. It is important to have a healthy and loving bond that brings the both of you together.

When training, use simple phrasing so that they can better understand what you are trying to say. Don't have complicated commands and explaining things in sentences doesn't help them. Simplify your talk. While we all love talking in long winded sentences to our dogs and they like to hear from us too, they don't really understand every single word except for relevant command words.

Ensure that you reward good behavior immediately after it happens.

Don't wait too long to do this. Right after they've done something you approve of, show them immediately.

Be patient in this process. Don't overwork them. They have incredible power within their brains and can be very smart creatures. At the same time, they do have a mental limit on the day, and it is probably less than what we have. Don't expect to train them all day. Include breaks to ensure that the two of you focus on playing and relaxing together as well.

After they've started showing signs of good behavior on their own, you can reduce using treats to reward them. During the first few weeks, you'll probably need to use treats frequently to get them to sit, lay down, go outside, and so on. Once they are doing these things regularly, cut back on the treats that they have. Dogs older than eight months won't need a treat every time they go to the bathroom. Too many treats could mean that they gain weight and end up being at risk for other health conditions.

Act excited when they are behaving well instead of using treats every time. You could still give them treats every time if that's something you want. Just ensure it is a healthier treat like frozen peas or carrot sticks.

Make sure to always finish up training sessions with a positive reinforcement. Even if you trained for 30 minutes and got nothing accomplished, end it by doing something that you know they're good at, like simply sitting. This ensures that they still feel good about training and that they'll be able to pick up where you left off next time you start again.

Studies have proven that the more engaged you are with your dog, the more obedient they are likely to be. If you take your dog on your runs and walks, let them ride in the car while you running some errands, and hang out with them while watching TV, they are more likely to be obedient rather than if you leave them at home or away in their crate during these times of activity.

Specific Rules for Small Dogs

Small dogs are great companions. They have small waste, they're always by your side, and they usually love to snuggle up right next to you. They are different from other dogs, so it is important to consider the ways that we should train them especially carefully to ensure their needs are being met.

What we have to remember is that the same rules apply for these smaller dogs, but we also have to take their size into account. They might have less energy to train for longer periods of time. They might be more passive and scared in larger settings because they are so tiny, especially as puppies.

Just because a Chihuahua might not hurt you when they bite you, that doesn't mean we shouldn't stop that behavior. We have to have the same rules for our small dogs as we do our big dogs. Even if their feces may be small enough to the point that we don't care if they go in the house, we still need to discipline them when they soil a place that isn't meant for them. However, if you have bad habits in one area, it can make more bad habits in other places.

Your smaller dog might initiate a fight with a larger dog, and that's when things will get dangerous. Though you don't care if your tiny Pomeranian is aggressive, it still teaches them that it is okay to be more aggressive. Then when they run into a Doberman on the street, they park and snarl, and the bigger dog fights back. This is where you would see the negative effects of improper training or condoning aggressive and negative behavior, so don't wait until things are heated to take action.

Train small dogs to do simple tricks when you are sitting down rather than standing above them. They can be easily intimidated by size, so make yourself as small as possible when working with them to teach them new tricks.

Don't assume that you should always be picking them up either. It is easy to just lift them up and plop them outside really quick rather than harnessing them up and walking them down the stairs. However, they need to learn the path to important places on their own, so walking them through rather than carrying is the best way for them to learn.

If you have a small dog and a big dog, make sure the small dog doesn't have more privileges than the bigger dog. You might let a Mini Poodle sleep in bed with you, but the German Shepherd has to sleep on the floor. This creates imbalances in power and the small dog might think they have more control, leading to unequal treatments.

Don't assume just because you are bigger that you are the one in charge! Small dogs can be known for having the most attitude, so remember that it is important to still maintain your role as the leader of the pack.

Specific Rules for Medium Dogs

Medium dogs are usually breeds that might be family friendly. These are the dogs that the whole family can play with, hangout with, and have a bond with. Medium dogs are the standard for what we discuss throughout the book, so they don't have as many specific training needs. Rather than considering their size, it is a breed that you will want to cater to.

Specific Rules for Large Dogs

Big dogs think that they are small dogs in some situations! They don't realize how big they are. Just because they jump on you and knock you to the ground, they aren't trying to attack you. Many people associate aggression with big dogs, but it is mostly just because they aren't aware of their physical impact. A small dog might get excited and jump up on your legs when they first see you, and you might hardly notice. A big dog will do the same, but their leap could push you back or hurt your legs, automatically making them seem more aggressive. Remember, big dogs aren't intentionally being physically stronger! It is just part of who they are.

Big dogs are going to be high strung, more hyper, and of course more physically boisterous when they haven't been given proper time to exercise and play around. Since they have a lot of energy they need to expend on a daily basis, they need a lot of time to play and run around, and they need plenty of space to do this. If you can't provide them with consistent exercise, a smaller dog is a better choice.

Just because they're big doesn't mean they can't still get scared! Even if you are five feet tall and you are training a 100-pound Great Dane, they can still be very scared of you. Treat them with the same love and compassion that you would a small dog.

Bigger dogs might need more boundaries than smaller dogs, especially so that they don't scare other people. Even though you might not mind it if your Greyhound jumps up on you every time you get home, visitors, especially children, could be scared by this reaction. Always consider your dog's size and breed when training and interacting with them!

Chapter 4: Training Your Puppy Outside of Your Home

You may be able to rein in your dog at home, but in public, it is a whole other story. Not only will your puppy be more distracted, but so might you! Training should never stop, which is why this chapter will provide you with tips to ensure your dog is doing well in public. This chapter is filled with all the tips needed to not only carry out training in public, but to specifically know the best methods of training in new environments. It is healthy for your dog to socialize, so even though it might be harder in the end, you have to participate in these methods to ensure your dog is growing in the best way possible.

Going for Walks and Leash Training

Even if your dog has a big fenced-in yard to run freely in, it is still important that you train them to use a leash properly. You will be keeping them on a leash as you take it to the vet, groomer, to doggy daycare, and so on. It is better to have them know how to use a leash rather than always tugging and pulling away from you. Your dog needs to wear a collar in case it ever runs away or gets lost so that it can find its way back to you. Having a dog that knows how to be obedient on a leash also means that it will be obedient in other important areas as well.

As soon as you bring the puppy home, get them used to wearing a collar. They don't necessarily have to wear it all day throughout the house, however, they should get used to wearing it at important times such as going outdoors. They might not like it at first, but eventually, they won't even notice it or feel it. It is important to get one that is the right size and strength for your dog. This is something that you will want to spend the most money on compared to other products within this category that might not be as important. You don't want to get a cheap collar that breaks easily!

A collar is necessary to help identify the dog, should it ever get away. There will always be a level of unpredictability when you have a dog, no matter how responsible you are as an owner. Your dog might be completely relaxed as you sort through your mailbox but then sees a squirrel for the first time and runs away as fast as possible, completely shocking you as it gets away! It can happen, so a collar and leash are important no matter the temperament of your dog.

Remember, your dog will pull and tug at the leash, trying to walk as fast as possible. Retractable leashes are banned in some areas, so your choice of leash type depends on where you are located. A leash that is at least five or six feet long is good for the average person and dog. Longer leashes are better for small dogs and taller people. The leash needs to be strong enough to keep the dog back. Train your puppy in a way that teaches them not to pull the leash as hard. If you have a very large and strong dog, a chain leash might even be something that you consider.

It is certainly beneficial to use a harness while training your dog as well. This way, you can pull the dog back without hurting them. There are a few different collars that you might use aside from just a harness too. The most common is a simple collar with a plastic buckle. This snaps together comfortably for your dog.

Another popular kind is the **martingale collar**. This is for dogs that have the same size neck as they do their head. It will be easier to adjust these so that it doesn't slip off.

There are also **half-choke collars**, which are looser collars that will tighten as the dog starts to pull. This is to keep them safe while teaching them not to pull as hard.

Prong collars have pieces of metal that stick out so when the dog pulls on the leash, the prongs poke them and make them uncomfortable. This is beneficial for the owner and training purposes, but it does hurt the dog. Some dogs will be fine and stop pulling when they feel the poke. Others won't even care and will continue to tug the leash as hard as they can. Consider your dog and their level of curiosity before you use a prong collar.

You can also get a collar that wraps around their face, known as headcollars. These keep their mouths closed as to encourage them to not bark as much. It is still attached to their neck so that you aren't tugging them by their face. It is best for dogs that bark frequently.

It is helpful to have a harness connected to the collar so that the dog knows they need to heel. While you don't want to choke them out, a little discomfort can be enough to teach them how to properly walk next to you. You might just think that dogs can easily start to walk as soon as you put the leash on them. There are a few important steps when introducing walks to your dog.

Book 3 - Training Your Puppy Step-By-Step

Stick to one side when walking. Whether you want them to walk on your right side or your left side is up to you. However, try to pick just one that they stick to in order to make it easier for them to be obedient when you start to go for walks.

Play with them first to make sure that they are nice and tired. If you go for a long walk first thing in the morning, they will most likely be active and excited to see everything that's around them. It is better to make sure that they've been playing awhile and are more tired when it comes time to take them for walks.

It is important to start indoors so they can get a feel for what's ahead. Once you are ready to start training your puppy, put a leash and collar on them. When they are sitting next to you with their collar on, this is a good time to reward them. Give them a treat for sitting right next to you and not moving. Start walking from one point to the next. When they take the lead, stop and wait until they are sitting right next to you. Feed them a treat again.

If they pull, do not pull back hard. Simply stop so that they know to stop, make them sit, and walk ahead a bit so that you are leading. Feed them a treat and use verbal positivity when they act this way. It is okay if they get a little ahead of you, but never to the point that they are tugging on their leash. They should learn how to walk right next to your side in order to make sure they are obedient when going for walks outside.

Be happy and engaging with your dog. Don't get overly excited as this might cause them to be too excited and then they'll want to run ahead of you. Practice walking indoors as frequently as you can. Use a clicker instead of treats as you transition to walking outdoors. If they are really struggling to not pull and don't want to walk next to you, then you will have to take treats outdoors. Remind them that you have a treat so that they'll be more interested in you rather than interested in what's ahead. If you are exciting, your dog will be focused on you rather than letting their attention carry them everywhere outside.

Doggy Daycare

Doggy daycare is a great way to start to make sure that your pet is well taken care of. You can find experts that love dogs to help keep your pets happy all day long while you are at work. Not only is it helpful for you, but it is better for your puppy. They'll get a new and exciting environment, interact with other dogs, and come home tired, so that you don't have to exhaust them right away.

First, do extensive research into your doggy daycare. Make sure to read real reviews from actual people from multiple sources as well. Do your research on what they practice, what they believe, and the basic steps they take their pets on throughout the day.

Consider other amenities that will benefit your dog the most. Do they offer grooming and bathing? Do they make sure to train dogs? Do they feed them? Do they have indoor and outdoor space to run? All of which are important things to consider for your pets. They should have a thriving environment that stimulates them and fulfills their basic needs.

Ask if you can tour the place first! Many will be fine with giving you an interview opportunity. They might not show you where they keep the dog out of respect for other dogs and their owners, however, it is beneficial to examine what the kennel situation looks like for your pets.

They'll likely have a training period with your dog. This might mean that they try things out for a week to see if it is a good fit or if your dog should go elsewhere. Socialize your dog at parks and other public places first to see what their strengths and weaknesses are. Do they get excited and run up to other dogs? Does this interaction make them anxious? Aggressive? Overly excited? Keep track of how they interact so that you can help the daycare workers as well.

Check in with what foods they'll have. Do they provide food or are you going to supply it? Will they be doing feeding at all?

Book 3 - Training Your Puppy Step-By-Step

Make them aware of all your commands. You want to share with them the tricks your dogs know or are at least trying to learn so that they can help your pets training further. You don't want to give them two different commands for the same action because that will just confuse them.

It is a good idea to find a daycare that also does training. This way, your dog will have consistency. You can take training classes with the dog and they can have the opportunity to learn from professionals, and then you can be assured all training procedures will be the same at home and at daycare.

Make drop-offs quick and don't get too excited when you see them for the first time. if you prolong the drop off, it will make them more anxious and scared. It will make each drop-off harder and harder. Make it short and sweet. When you pick them up, they will naturally be excited. Don't make it so that they are even more out of control with their emotions. It is tempting to be like, "HI! I missed you so much!" with a screeching, excited tone, giving them lots of hugs and kisses. How could any dog parents resist? However, this will make these moments harder and harder each time because the dog will only associate positivity with you, and not with the trainers or caretakers they're with all day.

Parks

Parks are like a dog heaven. Whether it is a regular park or one that's specific for dogs, they provide a great opportunity to get some exercise while healthily socializing as well. Find a park that is close to your home which you can take your dog to regularly. It is good to find one in walking distance so that you can get more exercise for your dog before they even get there.

Check-in with all the rules allowed at the park. Are dogs allowed? Are they allowed off their leashes? You don't want to take your dog somewhere that they aren't even allowed! Remember that they will be socializing with other dogs and other people. Are they prepared for this? Do they have the skills necessary to be a present animal that doesn't attack others? Will they get excited and jump on someone? Will they bark and growl at other dogs? While it is important to socialize your pet, you also have to remember that there needs to be a level of basic training already instilled in them so you can diffuse any potentially challenging scenarios.

Consider your dog's anxiety and excitement level. Are they going to be freaking out and be overly excited? Are they going to be more relaxed and instead stay close to you?

Walk around with them before going to the park to reduce their excitement. If you take them to the park without having any physical exercise in the day yet, then they will be more excited than anything else. Don't wait until you get to the park to work them out. They need time to calm down before playing or else they might end up being far too excited.

If it is a specific dog park, then it is usually fenced in so they can be off leash. Other parks might have fences that allow you to take them off their leash as well but check in with the rules. You might also want to take them off their leash no matter what, but you have to ensure they are properly trained before ever doing this. If you ever take them off their leash with no fence, they have to know how to come on command, how to walk next to you rather than ahead of you, and their excitement with other people should be low. Even if this is the case, it is still not recommended to let them run free where there's no fence. It is not just for your sake, but for other people as well.

Don't take their leash off right away when you get to the fenced in park either. Give them a chance to sniff around. Let other dogs come up to them if they want and give them a chance to look around and get used to their surroundings. If you let them off their leash right away, they might start darting around and become concerned and fixated on the wrong things. Warm them up to this new public setting by keeping them on their leash for the first five minutes, minimum.

Don't build the excitement too much either. It is easy to say things like, "Do you want to go to the park?" in an excited tone. How could we resist seeing and showing off how cute our pups are when they get all excited. However, this is going to make them more active when you get to the park. They could be excited to the point that they're uncontrollable, and that's not what you want for your dog.

If you are taking a puppy that's not yet spayed or neutered, consider how other dogs may react. If you take your tiny Mini Poodle girl to the park and she's not fixed, she might get approached by a lot of larger male dogs. This could cause her anxiety, and she might even fight some back. Don't let this happen and keep her close if that's the case. If you have an unfixed male, he might run up to other females and try to bother them as well. It is best to have all adults fixed, but for puppies that are too young, keep them close.

Feed them long before so that they can go to the bathroom before you enter. If you feed them right before you go, they might poop right away when you get there. That's fine for dog parks, but for regular parks, this could be against the rules.

Pay attention to your dog and interact with them constantly. This shouldn't be the time to scroll on your phone. You never know what they might get into, and you will know even less what other dogs could end up doing. We can't assume that all dogs at the park have been properly trained in the first place.

It might be a good idea to leave the toys at home. A ball is usually fine, but specific toys could cause conflict. Another dog might approach your puppy, and they could get defensive over the toy. Dogs tend to be more competitive at parks. Remove the chance of conflict by leaving potentially triggering items at home.

Pay attention to what they might be picking up and sniffing at. Especially puppies. They might stumble upon a fresh pile of feces left by another dog and decided to give it a lick to see if it is a treat. The point is to have fun, so don't make it a stressful time for your dog! Be as prepared as possible to meet all their needs.

Friends' Homes and New Environments

There are a lot of people in the world who don't care about how other dogs act. They'll be excited to see a puppy and the first thing they'll want to do is shower it with attention, even if your dog might be misbehaving. They won't always be as serious about training because they aren't thinking of the dog's needs in the long run, they simply want to hold and cuddle a cute little puppy.

Though you might not care if your dog is jumping on strangers, remember that going over to someone's house might mean that they'd prefer if you didn't allow your dog to do that. You might have no boundaries for your furniture and the dogs are allowed to sit and chill whenever they want. Other people might not want your dog on their couch at all, so respect the rules of other people's homes.

You might have your friend be pet sitting as well. Whatever the situation, there are a few rules to remember. Keep them leashed when you first arrive. Just like the dog park, they are going to be overly excited, so you need to give them some time to warm up to their new surroundings. If you don't keep them leashed, they might end up running rampant around the new home to sniff out everything and anything!

Have a toy that will keep them distracted. If you are going to a friend's house to watch a movie or play a game, you'll want your dog to be focused on something other than you. Bring a bone for them to play with or another treat that will keep them busy on their own. This will help make sure that they are relaxed and free from getting into typical doggy trouble.

If they're staying the night, ensure that you have a place for them to sleep. Bring their crate with you if they are still crate training. You want to ensure the same consistency in night time routines at someone else's house that you have for your own home.

Check-in with any other pets in the house and what it might mean for all the animals. Does your friend have cats at his or her house? Has their cat interacted with dogs before? t They might need to keep the cats in a bedroom during the visit to protect both the dog and the cat.

If you are introducing it to a new dog, have your friend bring them outside. If you bring your puppy into a dog's home, this could make them defensive. Instead, have them meet on common ground so that they're more respectful of each other, and then you can bring them together.

Bring supplies in case they decide to use the bathroom in the house. While you might think you have it under control because your puppy hasn't gone inside for the past few weeks, this new environment could be the one place they decide to relieve themselves.

If you can, walk your dog through the house to the backyard or whatever the pathway might be to use the bathroom. Just like you would for your training at home, you want to show them the way to the bathroom so that they'll know exactly where to go when they need to relieve themselves.

Remember to be respectful of house rules as well. Even though you let your dog on the couch, your friend might not want that. This is why it is important to teach them 'down' commands when they're at home on the couch and bed. If they know how to get down, then you can use this command if they do end up jumping up.

What to Do When You are Not Home

You should take three or four days off when you first get a puppy. Unfortunately, there's no "new puppy" leave option for work. Imagine if we had puppy-leave like maternity-leave! After getting your dog introduced to the new home and when you have to head back to work, you'll need to use different methods to care for them when you are not home.

The first thing you should do is fill in help and reach out from friends and family to be with the dog. After six months, your dog will likely be in some healthy routines, though they're still technically a puppy. Before this time period, see if you have anyone, or a number of people, that are willing to come over and hangout or play with the puppy for a bit, or if they can at least stop in to take them outside fast. Many people will be more than willing to help out for puppies!

You can hire a dog sitter or take them to doggy daycare. There are many apps and other services that make it easy to find a dog walker in your area. They can send you pictures of the dog so that you can make sure they are being properly taken care of.

If none of these are an option, then you'll have to make sure to set the home up for them when you leave. Remember that they should never be in a crate for longer than eight hours, but this is the absolute maximum. That doesn't mean that eight hours and 26 minutes is okay. If anything, you should keep in mind that seven hours is your maximum threshold and you can go over that by a few minutes here and there, or on the days you are running late.

Take them outside before you go, but don't make it something that they are scared of. Treat it like normal. Leave your purse or other things you are taking with you inside so that they don't suspect anything. Just like with the doggy daycare, don't make them overly excited or else these will become focal points for your dog's excitement.

Wait around a few minutes after they go to the bathroom. If you take them inside immediately after, they will learn that peeing or pooping means it is time for you to leave. They might end up holding it for a while because they don't want you to leave. Then you might think that they don't have to go, so you take them back inside, only for them to have to hold it for even longer. They should have plenty of time to go to the bathroom.

Crating at first is fine, but only for shorter periods of time. Remember the month-to-hour ratio for how long they can hold their bladder. If you are going to be gone for a long time, you might consider giving them a room. This way they have several different options to lie around and play, and if they do have to go to the bathroom inside, they aren't doing it in their crate where they have to lay in it.

Tire them out if this is an option. If you can play with them before you leave for work in the morning, this makes it so that they can just focus on sleeping while you are gone. Invest in a doggy webcam so that you can watch what they're doing. Some even give you the option to press a button to talk to them or release a treat!

When you get home, the same thing goes for the doggy daycare. Don't make it too exciting or else this is how they react the first time they see anyone. Make it quick and casual and take them outside immediately. Remember that they have been laying around all day, so they will be rather energetic. After a long day of work, you may not be nearly as excited, so don't punish them because they're extra happy. Make it a point to play with them for at least 15 minutes when you get home from work, or wherever else you were. This ensures that they'll get tired out and you won't have to deal with a crazy energetic dog for the rest of the night.

Public Tips and Reminders

You might be getting a dog as a service animal. You might be someone who is more active than others. Whatever your reasoning, you will probably be taking your dog out in public frequently. We have a few more tips for you to remember when taking your dog out into the social world.

Keep your dog training even when in public. Consistency is key so you don't want to create an environment where your dog is free to do whatever they want just because you are no longer home. It will confuse them and make your training regress.

Just like the other activities we discussed, try to tire them out before going out in public. Establish some own boundaries with what you might say to those who want to pet your dog. You might not want others to distract them or make them excited, so it is perfectly fine to say, "I'm sorry, but we're in training right now," if someone asks if they can pet your dog.

Keep a few treats in your pocket so that you are not out of options should you have to reward good behavior. A clicker is a good option in public because it helps bring their focus back to where it should be without it being a distraction.

If they are acting badly, don't just give them a treat to calm them down. This will teach them that bad behavior will get rewarded. If you have one dog in training and one dog that's older and already well-behaved, try to refrain from having both of them out in public at the same time. This might make it harder to focus on training the one. Whatever you do, maintain consistency, and stick to the same rules no matter where you are.

Chapter 5: Skills that Dogs Need to Know

Though it is cute to teach your dog to sit and shake your hand, roll over or play dead, they also need to know some basic skills such as sitting and laying down. These are important things to teach your pup because it will be the way that you control them when needed as you are training. When you can teach your dog new tricks, they will be more obedient. Remember that these training processes aren't just to be done in the beginning, but all throughout your dog's life so that they're always effectively learning in the best way possible.

Sitting and Laying Down

You should start to teach your dog to sit as soon as they come home. Just because we're discussing this in chapter 5 doesn't mean you shouldn't start right away! Sitting is going to be one of the most important tricks that you will teach them. Even the most untrained dogs will still know how to properly sit. It is a natural state for them to fall into, which helps you have better control over them.

Remember positive reinforcement is key! Don't scrutinize them because they're not sitting. They won't understand why you are mad. Have some treats ready to hand out as you teach them this valuable skill.

Sitting is a great command because it is the way that you ensure your dog will be relaxed and chill when you need them to be. Maybe you are about to feed them and they keep running in circles around you. Perhaps new visitors are coming and you need them to be relaxed and sit down so they don't jump on others. You can command them to sit to ensure that they'll be as relaxed as possible in these times of great excitement.

Do these training sessions in shorter time periods. No more than 10 minutes is good. If you go for longer, it can tire them out. It can be too repetitive, and they might be desensitized, especially if they're not properly learning the different techniques.

To start, hold onto some treats. When they sit, give them a treat! It is as simple as that. You should use the normal command, 'sit,' rather than something more complicated. They might not sit on their own for a while, and that's fine. Simply give them a treat only once their furry bottoms are touching the floor and they're giving you attention.

After a few times, start to say 'sit' as they sit down. This is the way that you can associate the word with the action. Continue to tell them to 'sit' on a regular basis whenever you want them to be more composed. Each time they sit, say "good job" and give them a treat. You will discover that they know that if they sit something good will happen, so they'll do it more often. Ensure that you are also using positive reinforcement and that you have an overly happy tone.

Stick to this word because it is how others will know how to command them when you are not around.

Don't expect them to stand up again after sitting either. Step backwards or away to get them back on all fours, and then ask the command again. This is the way that you'll be able to teach them what 'sit' means rather than using your hands to force them to sit or to stand up.

Practice without the treats to make sure that they're getting it. Slowly phase out the treats after a couple weeks of this practice. They have to learn that not every time they sit they get a treat, but still use positive words so they know how good of a dog they're being.

They should learn to sit when:

- You are giving them something, like their food.
- Before you put their leash on to go outside.
- When someone arrives at your home.
- When they are being nosy and begging for your food.
- Before jumping onto beds/couches.
- Any other time you need their attention and focus.

Laying down is just as easy, but they need to know how to sit first. After you've successfully taught them to sit, you will be able to teach them so many more tricks even easier. It is like learning an instrument. Once you know how to play one, the rest will come simpler. To teach them to lay down, you'll have to start by paying attention to when they are actually laying down.

You'll do it in the same way that you taught them to sit. Wait until you see them plop down and laying there and give them a treat. Once they're laying down, give them praise. They'll likely stand up from excitement, so it is a good opportunity to practice over and over again.

As you see them start to lay down, tell them, 'down.' They will then know that this word will mean that they have to lay down on the floor. Another method to teach them to lay down is to put a treat on the ground with your hand over it. they will try to get as close to it as possible, so in the process, they end up laying down. Give them the treat a few moments after this so that they understand that it is the laying down which is giving them the treat, not that this is just a random occurrence.

Never force their back down. They won't understand what you are trying to do, and it won't be the way to teach them to sit. If you are just forcing their back down, they aren't learning how to do it on their own, so they will only do this when you are the one pressing on their back.

Bark Control

Barking can be a huge issue for dogs. This is one top reason that people end up taking their dogs back to shelters or give them away as well. To make sure that your dog's barking is under control, not just for your sake but for your neighbors as well, there are a few things you can do.

Dogs that herd and have other sports jobs might be more inclined to bark. They think that barking is the way of communicating and that it is part of their job. If barking is something that's going to be a big issue for you, it is best to first not pick a dog that is known for barking.

Look at the reason that they are barking so often. Is there a distraction that they always have to bark at? Maybe it is the mailman or other traffic that frequently passes by your house. Could it be that they are anxious and scared of something? Is your dog constantly barking because they think that it is playtime? Have they not been conditioned to stop barking long before this? If you look at the root of why they're barking so much, you might find a solution. For example, barking outside at neighbors walking by could be fixed if you get ticker blinds and block their view. If they're barking because they're anxious, you can do some things to alleviate those feelings.

Don't yell at them to stop barking. They will just think that you are barking with them. If they don't stop barking and all you do to stop it is add even more noise, the dogs will believe that what they're doing is right.

Wait for them to stop barking. It is frustrating but it is the best thing for them. After they've stopped barking, wait a few moments, and then give them a treat. You have to ignore them as they're barking or else they'll think about what they're doing is right. If you reward them when they're quiet, then this helps to make sure that they know quiet time is when you are the most positive.

Teaching them when to talk could actually help them be quieter. If you want to really control your dog's barking, start by teaching them to know how to 'speak.' To begin, wait for them to bark just once, and then give them a treat. Continue to do this and start to say 'speak' as they bark just as you did when you were training them to sit and to lay down. Rather than just barking whenever they want, it will help if they're trained so that they only speak on command.

This is an intermediate trick, so you probably won't really be able to do it right away, but it will still help with puppies since they can learn faster than older dogs. If training them to speak doesn't interest you, that's not the only way to have them stay quieter. Rather than yelling at them to stop, ask them to sit. When you ask them to sit, it snaps them into that mindset of needing to be more relaxed and quieter.

Remember that some dogs are just louder than others. There are certain devices that you can use if they bark a lot. You could purchase a noise maker that releases a high frequency pitch as they bark. The dog won't like the sound, so it might help them to stay quieter. You can use the clicker when they are good and click it as they are barking. It could help them to instantly snap into a calmer mindset.

Greeting Visitors and Walking Properly

As we mentioned earlier, when dogs meet new people, they will be highly excited. It is always fun to meet new people, and dogs don't have the kind of social anxiety that makes us as scared of other people! They will walk right up to anyone in most cases. The problem with this, however, is that not everyone likes dogs. Even those who are dog lovers aren't going to be as interested in having dogs jump all over them. It can be a bit too much when there is a big dog greeting you. The same goes for when you are walking down the street. A lot of people are going to be excited to see your dog and want to pet them, and your dog will notice their excitement, sometimes matching that and getting happy and jumping on them. There are a few things that can make this process easier.

Don't punish them for their enthusiasm. They are just happy to be meeting new people! They are simply excited that they are in a new environment and dogs aren't always the best at containing that kind of thrill! Rather than making them feel bad for being so happy, instead, make it your focus to simply remove them from the situation. Step back and make them sit or lay down. Once they have calmed down, give them a treat as a reward.

Continue to reward positive behavior. When they greet someone and immediately make sure to stay sitting, use positive reinforcement. Remind them that they are a good dog for not jumping on the other person.

Remember that this is a training period. You can tell your visitors this so that they are more understanding and they know that they might get caught in the cross-fire. For when visitors arrive at your home, keep the dog on a leash. Have them sit, and each time they get excited, make sure that they sit again until they are calm enough to have you let people come in. If someone knocks on the door and your dog starts getting excited and barking, make sure to not let the visitor in until they've calmed down. It is hard to not get your dog excited, especially if you are seeing someone for the first time after a while. You might want to tell them, "So and so is here!" so that they grow their excitement. Resist this so that they're as calm as possible when new people come in.

When you get home and your dog starts to jump on you, rather than stepping backwards, step forward. You want to remind them that you are in charge and you are the leader. They have to follow your lead, you aren't to follow theirs. Ignore them and don't greet them as they jump. Once they calmed down, this is when you can pet them, take them outside, give them a treat, and so on.

If they jump onto the couch and you want them down, touch the ground with a treat in your hand. Each time they jump up, do this, and have them sit on the ground. You can start to say to them, 'down,' so that they associate this instance with sitting peacefully on the floor. It will make it much easier for you and your family to tell them to not jump on them or on furniture when they have an actual command to get down.

As far as going for walks is handled, if someone comes up and asks if they can pet your dog, tell them that's fine, but make your dog sit down first. Remind the other person that they're in training and you don't want to teach them that jumping on strangers is acceptable. Having your dog know their name is going to be very important in this process as well!

Knowing Their Name and Coming on Command

Your dog's name is a very important part of their identity. This is their command. This is what they know themselves by. They associate this name with positivity, and it helps the two of you connect further.

Your dog knows to listen to their name because it means that you will be providing them with something afterwards. Calling them by their name from the beginning is important. Pick a name that is short. While Mr. Hashbrown Casserole the Third is an adorable name, you want something that is easy to call them. If you do have a long full name, that's fine, but have a nickname that they are most known for going by.

Some individuals even think giving your dog a human name will make it harder for you to treat them like a dog and not a human. This is up to you, but keep in mind that naming them after a family member or friend, or with a name that you already know might make it harder to differentiate your emotional evaluation of your dog versus this person.

Don't confuse them with a word that sounds like a command. For example, you might name them Brownie, which could sound like 'down.' Mitt might sound like 'sit' and so on. Their name should be clear and unique to them so that in all situations they always know who you are trying to refer to.

Something with a 'c' or a 'k' can help your dog to better understand if you are saying their name or something else. It is a noise that separates their name from any other noises that they might be hearing.

If you adopt them, there's a good chance they already have a name. Many breeders will refrain from name-training them. It is perfectly fine to change up their name (Keliher, n.d.). As long as you don't call them by their old name anymore, it shouldn't be an issue when training.

Looking at them and saying their name is enough. Make eye contact and say their name. give them a treat after they respond in the beginning. Don't overuse their name or else you might confuse them or teach them that they don't need to pay attention. If they aren't listening after the first few times, take a break and try again later.

As far as calling them, ensure that you start by saying the name around them. Give them a treat whenever they come to you after you've said their name. Don't make them "come here" when you are angry at them and want to use a negative tone. If you use negativity around their name, they are going to reject it and be far less likely to actually listen to you.

Don't use their name to call them when you are giving them a bath, taking them to the vet, or doing something else that they don't like. This just teaches them that they should be afraid of their name and of coming to you when called. If you aren't nice when they come to you, then they simply won't want to come to you anymore.

Fun Tricks for Your Dog

Some individuals might think it is tedious to teach dogs how to do fun tricks just for our entertainment. It is actually good for them too! It keeps them focused, and when they're obedient, they will be better behaved in all situations.

You can start with a simple 'shake.' This is a cute way for them to interact with new people. Since they do it while sitting down, they'll likely be more obedient as well. Have them sit down. You can simply wait until he puts his paw in your hand and then reward.

Book 3 - Training Your Puppy Step-By-Step

You could also try to tap the back of their leg to get them to put their hand up. Don't grab their paw, however. It is the same as trying to force their back down when teaching them to sit. They have to learn how to do this action on their own. It should be a natural motion. It will take some time and you have to be patient, but they will learn how to do it.

For rolling over, have them lie down. Hold a treat behind their backs and over their shoulder so that they have to roll over in order to sniff the treat. Get them to lie on their back, and then have them roll back into a lay down position. Once they completed the roll, give them a treat.

Use a treat to have them roll onto their back over and over again, never giving them the treat until they've completed the entire roll. Once they've done this, it becomes easier for them to understand what it is that they're doing which gave them the treat as a reward. Do these tricks daily in order to get them to understand the different steps they have to take to complete the trick for the reward.

Once they complete a roll, use verbal praise as well. Treats help of course, but it is your enthusiasm that is also going to make sure that they understand what they're doing is right. They will always be looking for your approval!

Having them "play dead" is super fun and will help them be the entertainment center of the party! They will love the attention. It is the same trick as rolling over, but instead of finishing the whole twist, they'll stop when lying on their back.

For this, you can associate a hand signal. Some people will hold their fingers out like they shot the dog, making it lay flat on its back. Others might ask a funny question, like, "Would you rather be a stinky cat or be dead?" This part will be up to you! It can be fun to play around with your command.

Reward them when they are in the position of playing dead. Use the treat in your hand when you make the hand gesture. This way, they'll know that the hand signal is what keeps them laying down.

Don't give them a treat right away or else they will think they have to snap out of this position quickly. Have them lie there for a moment and then give them the treat.

Practice these daily with your dog. They won't learn right away, and it will take some time for them to understand. Alternate between giving them a treat and doing it with no treat so that they do it for positive praise and not just a snack. If you give them a treat every single time they do it, and then just stop giving them treats altogether, they'll eventually stop doing the trick because it no longer means they get any treats. Don't give up with your dog, remain patient, and always remember positivity!

Chapter 6: Kinds of Exercises for Your Puppy

Playing with your dog and taking them out to the bathroom is important, but they also need to have specific moments of exercise. Just like humans, dogs can be more at risk when they aren't getting the healthy exercise needed. It could be bad for their joints, their hearts, and their bodies in general. Don't wait to exercise them! Start healthy habits from the beginning. You might discover that it actually helps you be more active as well!

Training and Agility Classes

A good method of exercise is to actually have them work with a personal trainer. If you have the means in your budget to hire someone to train your puppy, then this is going to benefit them and you. You can work alongside the trainer to learn important tricks as well as breed-specific training to help them be the best dogs they possibly can. Training isn't just for disobedient dogs – any of them could benefit! It is not even something you have to do right away either. Consider this as an option when deciding what kinds of exercises for your puppy to use.

Not everyone can afford this, so you can start to use training and agility cues at home. These are courses designed to test your dog's mental and physical skills. It is healthy for them because it provides them with exercises, and you have the ability to work with them directly, strengthening your bond. Your dog will be excited to complete tasks while also getting a fun workout that will improve his body health.

It is important that humans keep a critical mind so that they can function higher, but this isn't just true for us. Dogs with a critical mind will be more obedient. They will feel more connected to you. When they have more agility training, if you command them to do other things, they will be more likely to respond.

Agility is the kind of sport that dogs participate in where they run through hoops, fences, and other obstacles. You can purchase courses for them or make them on your own. Some individuals will use PVC pipes or pool noodles to create courses. There are common ones that you can try to recreate, or you could also come up with your own.

Your dog's safety is the most important part. Ensure any homemade courses are safe and that the dog's health and safety isn't at risk. It can help alleviate anxiety and make them more tired each day when you do consistent agility training. We just have to ensure we aren't pushing our dogs too hard when they might be feeling tired.

Start simply with some hurdles to jump over. These can be purchased at most athletic stores or you could use your own material. From there, you can set up cones for them to weave through.

To train them through these courses, you can use a simple treat in your hand to lead them through, or a longer stick with a treat at the end to help them go faster. After going through the course several times, a day for a couple of weeks, they will know how to run through with a simple command. There are many videos on YouTube that can help you deeper understanding how to train your dog for agility skills. It should be fun for them and time to play with you, not a moment where they feel stressed and scared of performing.

Swimming

Swimming will be a great exercise for your dog to participate in. It is easier on their joints and gives them something simple to do. It is a great way to cool them off in the summer as well. They will be happy to have a task, and this is another exercise that can strengthen their bond with you. They will trust you to care for them in this time.

A lot of dogs love swimming, but a lot of dogs also equally HATE it. It can cause anxiety and make them feel like they don't have control. It is a different sensation for them and not every dog can handle getting wet or feeling dirty.

A life jacket is a good idea for starting with certain breeds. Most will be fine, but there are some breeds, like bulldogs or pugs, that might sink right to the bottom when they are first in water.

Don't take your dog out to deep water where there might be dangerous currents. Keep it on the shore in the beginning to make sure that their safety is entirely protected.

Make sure that they aren't drinking the water they're swimming in. Lake water can sometimes carry certain bacteria harmful for your dog, and pool water has harmful chemicals for your puppy.

Introduce them slowly. Tossing them in the water can be traumatizing. They just might never go back to water again if you do this! Ease them into it so that they can be comfortable with the water around them.

Dogs don't like to be dirty. It is good to go for a walk with them where they can have their legs be wet while still being in control of their body. If you are wanting them to go swimming in a body of water, have them walk along the coast with you. Have treats with you so that you can give them positive reinforcement when they let the water touch their feet. Don't force them in just yet and guide them along where the land meets the water. Let them sniff and pat their paw in the water. They'll be curious so encourage that in them!

If you are taking them into a pool, buy a kiddie pool first and fill it up with just a few inches of water. Throw a ball in or a different toy they like so they can retrieve it and feel the water on their feet. Let them splash around in this safe area before you pull them into the actual pool.

Give them the option to leave if they are feeling uncomfortable. You should keep them on a leash since they might even run in fear but let the leash stay loose enough, so they have plenty of space to run away if they get scared.

Never command them to go into the water and only use praise and excitement. Don't tug them into it by their collar either. It should be a natural thing that they decide to do on their own.

Once they're in the water and floating, they'll naturally try to keep themselves afloat. Don't assume all dogs can swim right away! You'll have to be there to help them stay afloat as well. Support their stomach. So that they stay close to the top of the surface. If they start to sink and their face goes under, it might scare them to the point that they freak out and try to frantically swim.

Guide them to the edge or the surface if they are scared at any point. They trust you, so don't make them break that trust by forcing them to stay down under the water! Wait until after you've gotten out of the pool and rinsed off to give any treats.

Always supervise your dog around water. Even though they might be a great swimmer after a while, they could still slip into a pool or get lost in waves from natural bodies of water.

Catch

Catch is a great way to exercise your dog, one won't always require you to be running around just as fast. It is good to run with your dog, but it can also be exhausting! Playing catch gives you the chance to relax and enjoy the outdoors while your pup gets plenty of exercise.

A lot of dogs will naturally catch. It is in their biology to want to retrieve things for their leaders. They know that you enjoy the ball, therefore when you throw it, they'll chase after it.

Start with a toy that they will actually want to play with. Not all dogs are that interested in balls. If they prefer a rope or other rubber toy, throw this for them. After you've done this a few times, you could transition to a ball.

Reward them as soon as they start to play with the toy. Use verbal praise to remind them that you are excited and that what they're doing is right.

Put it on the ground in front of you and encourage them to play. This primes them so that when they catch the ball, they bring it back to you rather than run away with it. Once they bring it back to you, you can then give them another treat.

This is a good time to teach them to "drop it" too. Many dogs will bring the ball back, but they won't always let it go. They can't eat a treat with a ball in their mouth, so you can tell them to "drop it" and when they do, give them a treat. As they're eating you can take the ball from them. This is a great method to help them learn this trick. Next time they pick up something they're not supposed to as well, you can tell them to "drop it" and they'll know what this command is.

Dog-Led Walks

Walking can be so routine for dogs. It is a way that they go to the bathroom, get a reward, then go back inside. Most of the time, they take your lead. As a fun exercise to do with your dog, let them be the ones to decide which way you are going to go. Some individuals believe that you shouldn't let the dog lead because this is destroying your role as the alpha. That is debatable, so it is up to you to decide if this is a method you want to incorporate or not.

Every once in a while, you can still let them take the lead. Don't let them pull on it, but if they want to go to the left or the right, backwards or forwards, let them be the ones to decide. Keep a loose leash so that there isn't as much strain as they walk around.

This is best in nature areas rather than structured streets where they might be passing a bunch of other people. There will be fewer distractions so it will be more about them exploring their surroundings rather than getting excited about new people.

Go somewhere that you could take multiple paths. Give them options of how they can explore. It is fun and relaxing for your dog. It makes them feel natural and like they are accomplishing something.

Creating a Routine

The most important thing to do for your dog is to create a routine for it. Dogs need to be able to stick to a schedule. It gives them assurance that they'll be safe and protected.

Make sure that you are feeding at the same time every day. If they have a random feeding schedule, it will be stressful for them because they'll never really know whether or not they will have food to eat. They'll also worry that they won't be able to eat when they're hungry.

They will go to the bathroom around the same time as well. It should be a routine for them to wake up and go outside, go before bed, and a few similar timeframes throughout the day.

Sleep is essential for them. Some dogs can get anxious and not be able to fully sleep. If they don't have a schedule, they might not be able to become relaxed enough to get that deep sleep needed.

Most importantly, have an exercise routine. Not only will it be good for your pup, but you'll benefit from these periods of high activity as well. They will know the difference between play time and relax if you make sure that you are carving out specific times for them to exercise. Rather than having an active pup that acts all wild, you can have a dog that knows when it is time to lay down and when it is time to be crazy and play.

Make sure that you are also conscious of the methods that they might be associating place and specific locations with parts of their routine. For example, ensure that they have a proper place to sleep every night and that it is consistent. They will know when you take them to this place, that it is time to sleep. Have them eat in a similar place as well. If they are led to their food bowl, they'll know that it is time to eat.

Not every day is going to have the same schedule, but when you are doing those routine things, they should have the same structure each time. Keep procedures consistent so that there's no confusion on your dog's end.

Taking them to the bathroom first thing in the morning is a good habit. At the same time, it is ok to wait just a few minutes if you need to set up the coffee or go to the bathroom first yourself. You don't want to rush and stress your dog out with that first morning bathroom trip. Give yourself enough time so that this period can be calm and relaxing for your dog.

15 minutes a day is the minimum playtime you should have for ALL dogs. The bigger the dog, the more exercise they will need. If they have a day where they don't get out much, make sure that they're twice as active the next day.

Chapter 7: What to Do for the Health of Your Dog

Your dog has plenty of health needs just as humans do. Choose a vet in your area that you trust who also has good reviews. Though you might be able to train and exercise them properly, it is their health that is still important at the end of the day. Ensure they live a long and happy life by always checking in with their health.

Initial Shots

It is important that we take care of our pet's health by giving them regular vet checkups. Going to the vet can be expensive, but it is something that needs to be done. If you can't afford regular vet check-ups, it is probably best to wait until you have some money saved up to adopt a dog. When you do get your puppy, you should have a savings account with around $250 - $1,000 in it for any pet emergencies. You don't want your dog to get a random sickness or a serious injury that you wouldn't be able to afford to help them out in the end!

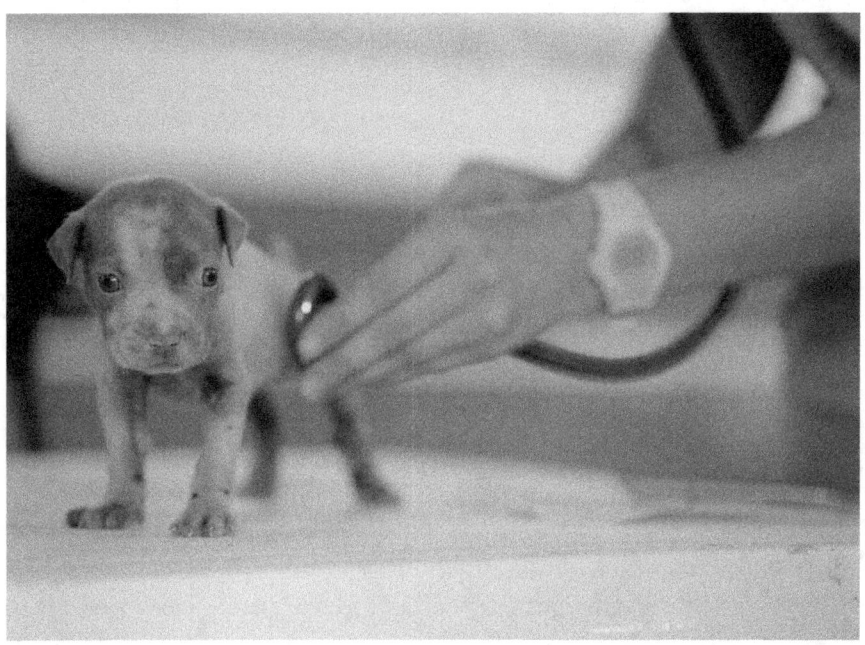

Many dogs will already have necessary shots if you adopt them from a shelter. This is a great option for those who can't afford all the shots up front. It is more convenient for you as well, so consider this before you adopt from a breeder.

Having vaccinations are also important so that when your dog is exposed to other animals, it can't catch, or spread any diseases. Puppies don't have the ability to get all the shots they need right away, so if your adult dog isn't vaccinated, it could affect your puppy. Aside from a few rare cases of having various reactions, getting your dog vaccinated isn't going to have any negative side effects.

Some laws require mandatory vaccinations. Check in with your local vet to see which ones you have to get in order to abide by your area's laws.

There are certain core vaccines that your dog will have to get. These are common health issues among dogs that you'll want to protect your pup from ever having to get. According to the ASPCA, these are the illnesses and diseases that core vaccinations will prevent:

- Canine parvovirus
- Distemper
- Canine hepatitis
- Rabies

There are other vaccinations that might be necessary for your dog. These could be things such as:

- Bordetella bronchiseptica
- Borrelia burgdorferi
- Leptospira bacteria

For puppies, the vaccinations should occur within six to eight weeks of their life. They will only get a few at a time with three-week periods in between (ASPCA, n.d.). By 6 months, they should have all of their proper vaccinations. Check-in with your vet to see if there are any that they'll have to come back for as they age.

It will be up to your vet and your dog's needs to determine what vaccinations are required. Its best to do as much as possible because the only downside is that they might cost more money. These vaccinations will often be a one-time thing, so it is not like you'll have to continually save up more money to keep up with regularly purchasing them.

Don't administer vaccinations on training days. They're going to be tired, and likely stressed out from going to the vet already! Give your dog a chance to rest and heal. The biggest issue you'll notice is that they are tired and that the injection site might be tender.

Anything too concerning, contact your vet. For the most part, however, they'll be completely fine and safe from getting anything serious.

Spay and Neutering

Not everyone will choose to spay and neuter their dog. You might use them to hunt or breed, so instead you'd prefer to keep all their reproductive systems intact. This is up to you, but unless you need their reproductive organs for a specific reason, you should be getting them spayed or neutered.

Spaying is when female dog's organs in their reproductive system are removed. Neutering is for males. For females, the ovaries and uterus are typically removed. For males, their testicles are removed. The actual penis and vulva are not removed on dogs. Emotionally speaking, it is a hard process for your dog to undergo, but in the end, they will be happier and healthier because of it. They might have a more relaxed temperament, gain weight easier, and be calmer in general.

It is important first to ensure that they don't reproduce. Having one puppy is hard enough, so you don't want to have more to take care of at once! After around six months, they will be able to start reproducing, so this is the age that they're usually spayed or neutered. Some studies have actually shown that waiting a year to spay or neuter larger breeds, like German Shepherds, is actually better for their health to give them the full chance to develop first (Dayton, 2016). Whatever you do, try to spay or neuter your dog between six months – 18 months of their life.

For female dogs, you will cut out their heat cycle. Some female dogs might even bleed when they are going through their heat cycles. It doesn't do much to them other than cause uncomfortability. They will likely be licking that area more when they're in heat, so they could end up causing sores if they do it too frequently. The bleeding could happen throughout the day as well, so it could end up staining your furniture or wherever else the dog might sit.

For male dogs, they might resist the urge to 'hump' things. Unfixed male dogs will usually do this to other animals as well as many toys. Even after they're fixed they might still do this, but it will be less frequent. It is usually not a reproductive issue and something that you have to train them to stop doing. Catch your dog before they start to hump. As you see them mounting something, command them to sit or lay down. Once they've successfully done this, reward them with a treat.

Both female and male dogs will also be more likely to listen, and it will be easier to train them. They will come on command easier and you won't have to worry about them being disobedient.

It can drastically reduce the risk of certain cancers or infections, such as pyometra. The lifespan of dogs that are fixed versus those that aren't is also longer.

You might be weary because you worry that it is not a natural process. It is also not natural to deprive them of these urges. It can end up hurting them in the end. They will still seek out other dogs to reproduce with and they might get stressed because they aren't able to find a mate. If they are fixed they will naturally be calmer, and they won't think about this as frequently.

Keep them away from other animals once you have gotten them fixed. They might be extra sensitive so you don't want to cause a fight that could make their stitches become loose.

Reduce their play time for the first few weeks after getting them fixed as well. You don't want to push them to overdo it or else they might end up hurting themselves in the process.

Give them two weeks to recover. This is the usual time period it takes for them to fully heal. They are getting internal surgery, so it can be an exhausting act to recover from.

A cone is probably going to be necessary because they'll want to lick their wounds. Make sure to keep this on them at all times that they are unsupervised, and even when you are together. It could only take one lick or bite to remove stitches, which could mean having to go back to the vet and do the whole stitching process over again.

Picking the Right Food and Treats

As you can see by now, the best method of training your dog is to use positive reinforcement and lots and lots of treats. For both food and treats, there are a few health guidelines that you should try to follow. Dogs aren't like us in that they won't care about tasty food as much. We have over 9,000 taste buds while dogs have far less, estimated at 1,700. They can still taste, it is just not as important to them. Many humans don't realize this and will often put their human emotions onto dogs. They think that these animals deserve to eat complex meals just as we enjoy, but this doesn't matter as much. We should be providing them with good and healthy food, but they aren't going to need to eat a steak dinner every time you do, so don't feel guilty that their food isn't as tasty. It is the smell of this food that they will care about more than anything (Finlay, 2017).

Dogs are also omnivores. Unlike cats, they should have some fruits or veggies in their diet. Raw veggies, like carrots, frozen peas, bananas, and apples are all great treat substitutes. Since you will be training them so frequently in the beginning and providing them with a ton of treats, you will want to consider these natural options to keep their weight under control.

The food you dog consumes will directly affect how they behave and how long they might live. We want our furry friends as healthy as possible so they can live a long time by our side! Choose the right food with added benefits and natural substances in order to ensure their health is in tip top shape.

Foods that are cheap and the popular brands that you can find at any convenience store will more than likely not be the first choice. Many dried foods are filled with carbohydrates that just add calories. Yes, dogs have to watch their carbs and calories too! Don't feed your dog only junk food. Many popular brands are equivalent to humans eating chips and other junk food as their sole meal.

The first three ingredients are the most important. Check any bag of food and look at the things that are listed before everything else. If these are chemicals that you can't even read, you might not want to have this be something that you give your dog.

The worst ingredients include:

- Meat by products (typically carcass or other animal parts that aren't actually meat)
- Caramel color (no dyes or additives should be in dog food)
- Oil (oil is dangerous for dogs and can cause inflammation)
- BHA or BHT (preservatives dangerous to health)
- Animal fat

The best ingredients will be:

- Meat
- Grains
- Vegetables
- Vitamins, minerals, and nutrients
- Meat meal

Your dog should be eating foods that have AAFCO approval. This means that it has actually been tested on dogs. Not all brands will give their food to dogs before even releasing it. Though it might taste good, it might not have any actual benefits for them.

Puppy specific food should be used for younger dogs. They have different nutritional needs that have to be attended to as they are still growing and developing. It will often be softer as well. Consider their breed, too. Some small dogs will have different food they need to eat, and larger dogs might have various needs as well. The tenderness and the size is important.

Wet food may be a better choice. The food that they eat in the wild is wet and moist, not crunchy bits all the time; however, you might like dry food because they don't eat it all at once. Check with the package to see what the serving size is. For puppies you will want to feed them three or four times a day. Once they're a year old, cut it back to just twice a day.

Water should be out all day long. If you have multiple pets, you should have multiple water bowls. It will be a good idea to have one more water bowl than people that you have. Never limit their access to water and ensure that no matter where you go, they will have the chance to drink as much as they want.

Avoiding People Food

It is hard to not give your dog a little snack when you are eating. You want them to like you and when you are enjoying food, you might want them to get the chance to enjoy this just as well! However, if you are giving them too many table scraps, it will create bad habits.

Not all people food is harmful to dogs, but you should avoid giving them your food when you are eating. It will create negative boundaries with your dog. It is a good idea to keep them in their kennel when you are eating so that they aren't trying to bother you too much while you have food in front of you. You should feed them after you eat, as this indicates to them that you are the leader of the pack, since in the wild, the leaders get to eat first.

Avoid giving them any human meat because there's so many factors that could be harmful. Even if you get it at a restaurant and want to give them scraps, you never know what they might have seasoned it with initially.

Fruits and veggies are fine snacks. Give them to your puppy as you wish but remember to not do it in the space that you eat. They should know to respect your boundaries and leave you alone as you try to eat your meals. These healthy fruits and veggies include:

- Carrots

- Bananas
- Apples
- Watermelon
- Cucumbers
- Peas
- Sweet potatoes
- Celery

These are great treat alternatives. If you don't want to buy a ton of treats all the time or find that they cause a ton of weight gain, then switching to these foods is a better option.

Make sure that anything you give them isn't seasoned. Dogs can't have garlic and onion, among other flavors, so you want to ensure that there are no hidden seasonings on the snacks you might be giving them.

Make sure that you never give them scraps directly at the table. They will know that it is fine to beg. Take the scraps to where they eat. Create a habit of making them sit or lay down in a different room when you are feeding them. You might think it is fine to do this at your home, but when you go to a friend's home or have entertaining guests over some meals, you don't want your dog to constantly bug them.

Ask yourself if your dog would find it out in the wild to eat on their own. They might naturally find certain fruits and veggies, but popcorn, chips, and bread they wouldn't. The key is to give them things that are going to help their health, not just satisfy their taste buds. If your dog doesn't eat their dog food, then you'll have to consider if they simply don't like it and purchase a new brand.

Flea Treatments and Bathing

Fleas are annoying, but they won't be the worst thing that can happen and are usually easily treatable. The worst thing that you will find if you get fleas in your home is that you get bitten and your bites might be itchy and swollen. This is treatable with some Neosporin ointment or other anti-itch cream to help your bug bites. It is worse for your animals, however, as the fleas can actually feed, and breed off them. Not only will the fleas bug your pets, but they will scratch, and this could cause even bigger sores and tender spots on their bodies.

They will get trapped in their fur and it can cause your dog to aggressively itch themselves. We scratch our wounds as humans, but we usually know when we need to stop before it gets bloody. Your dog doesn't know as well and will scratch and scratch and scratch. This can make their wounds even bigger and cause infections. If you notice your dog frequently scratching, it might be a sign that they do have some sort of insect issue.

Though fleas are more treatable than other issues, don't pretend like they're no big deal, because they can still carry viruses. Some fleas can go from two to 2,000 within a couple of weeks!

You can look for fleas yourself by separating their fur and looking at the root of the hair. On their skin you will be able to see the flea crawling around, or you will notice the flea dirt.

When you do see a flea, the only way to ensure it dies is to drown it, as squishing them doesn't always work. They have a hard-outer shell, so it is often the best choice to flush them or rinse them down the drain with soap and rubbing alcohol. If you do find a bug and it is not very tough to squish and kill, then it is probably not actually a flea and instead some other sort of insect.

You can give your dog a flea bath and then give it a flea treatment. The first bath will be to make sure that as many fleas and flea dirt is washed away as possible. You'll then want to give them a flea treatment. The best kind is an ointment you put on the back of their neck. This soaks into their natural bodily oils and will eventually spread throughout their entire body. About a half a week to an entire seven days from giving the flea treatment, you could do another bath to get rid of any fleas that might have died. Check the box of the flea treatment you purchase to see if you need to wait longer.

Treat your dog not just for its own sake but for other pets around the house as well. Your dog that goes outdoors might bring home fleas to your indoor cats. If you have indoor cats and a dog, you will still want to give them treatments. The dog might spread fleas to them, and then the fleas could spread back to your dogs. It is best if simply all cats have been treated.

If your home seems to have fleas pretty bad, you can do a home treatment. This is usually in the form of a smoke bomb that will spread throughout the entire house. Ensure that no animals or humans are home when you do these treatments.

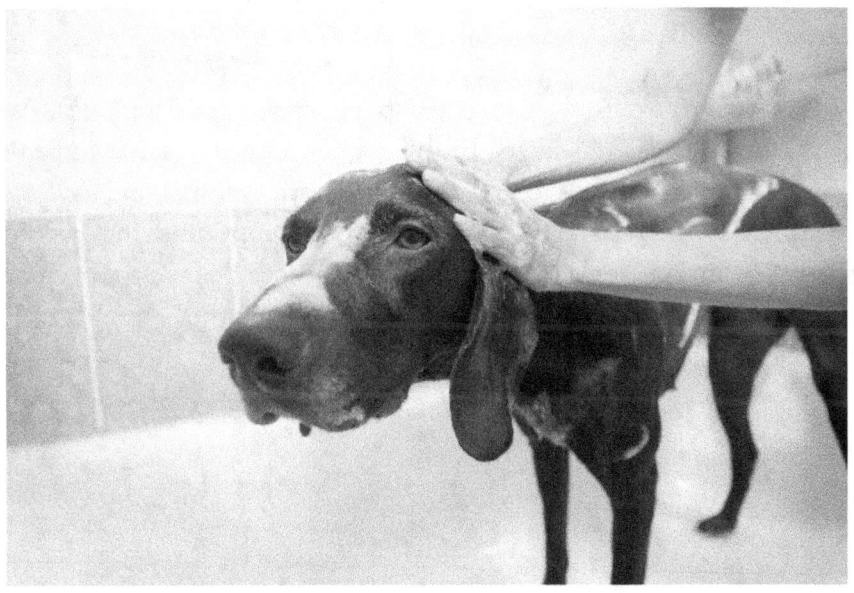

As far as regular bathing goes, this will vary depending on the dog. Some dogs will need regular bathing while other dogs will be perfectly fine going a year or more without a bath! It will depend on the dog breed whether or not it needs bathed.

Short-haired dogs will need bathed less. Make sure that you regularly brush all dogs, short or long haired. They will still need brushed in order to remove dead skin and loose hairs. It will help regulate the hair they leave behind on your furniture as well. Outdoor dogs that are very active need more bathing as well.

Once a season, or around three-or-four-month intervals is usually the minimum for bathing. This would be for bigger short-hair dogs who don't need as much grooming. The most frequent you should do is every other week. This would be for dogs that need regular grooming and whose hair grows long.

Keep their hair groomed so that it doesn't get too long between their toes or too long over their eyes. If you live in a hot area, it is best to maintain a buzz cut so that they don't get too hot. When it is cold outside, having a thick coat will help them stay warmer and make it easier to encourage them to go outside.

When bathing, pick only a dog-shampoo. Sometimes people will use baby soap, but there might still be dyes or perfume in this that could be fine for a human but not for a dog. Make the water warm, and have the environment be relaxing. Try to ease your puppy into the bathing process by making it exciting and giving them treats. Some people have actually used the trick of spreading peanut butter on the bathtub wall so that the dog licks it while they're getting bathed!

Chapter 8: What Not to Do When Training Dogs

There are some common mistakes in dog training that you'll want to do your best to avoid. There are many dog owners in the world that might not give their puppies the best chance when training, and then they end up wondering why their dog isn't behaving the way that they want it to. To make sure that your dog is going to be properly trained, we've created a list of what not to do so that you don't make any of the common mistakes that other pet parents might!

Aggression in Training

As we already discussed, aggression will just make them react more aggressively in the end. Though they might get frustrating at times when they don't listen, and they might have times where they don't have any progress in training, using aggression on them can cause them to be afraid and will actually regress any valuable training you did use. These are the most common forms of aggression and you shouldn't be using it on your dogs ever:

- Hitting
- Kicking
- Forceful pushing
- Threatening body stance
- Verbal aggression
- Yelling
- Pointing
- Other forms of physical and verbal violence

Some dogs are rather aggressive, so you might ask yourself, "What am I supposed to do to control my dog?" You might have a 120-pound dog that never seems to listen when you are telling it to sit down. Perhaps you have a little poodle that never stops using the bathroom in the house. Though these are frustrating situations, violence isn't going to help. We're going to discuss a few ways that you can assert power that you don't need to use any violence on them.

Not only does hitting your dog hurt them physically, but it destroys that bond. You are the one person that they trust more than anyone. How are they supposed to trust you if you hurt them so much? They won't be able to listen to you because they're scared of you. The more aggressive and angrier you are with them, the more that they will have trouble following your lead. They will grow their fear, not their trust, and that's a quick way to kill anything the two of you have built together.

Let's first understand the reason that they're aggressive. Dogs are very complex creatures and they have some needs that have to be met, while also having different stressors and anxieties just as humans do. Though they don't feel the same way that we do, they can still experience some of the same effects of stress.

The first reason is because they are trying to be protective. Dogs are territorial animals, like many other creatures. They want to be able to help protect your home. They care about you and want to ensure that you are protected. If they are often aggressive towards others, they are likely trying to protect their turf.

They are often frustrated and feel the need to gain control in some way. Just like how some people are more easily flustered, many dogs can be confused and turn angry too. They might be overwhelmed by others and feel the need to be aggressive to try and gain control. When you don't know what's going on, you might find that you panic, and this is how some dogs will operate.

They might be aggressive because they haven't been socialized. This is why it is important that we start to socialize our puppies from the moment that they're brought home. If we don't, we could end up making them scared of visitors and more likely to lash out on others.

Animals have a fight or flight instinct just like humans, so showing signs of being aggressive might simply be because they are anxious. In order to assert dominance, there are a few things you can do.

Be consistent with the boundaries you set. You can't have rules that apply during one part of the day, but not the other. You might not care if they jump up on you, but you have to set this boundary so that they don't jump up on strangers.

Sometimes, forcing them to be affectionate can actually damage that bond. If you are constantly squeezing and kissing them, or even holding larger dogs that don't like to be picked up, it could make them more fearful. They might not know what you are doing, so let the dog come to you when you want to show your affection. Be gentle and patient with them. If they are sitting next to you and licking you, you could try to put your arm around them and be affectionate, but if they're running around, eating, or playing, don't try to force a hug.

You have to have confidence over them. They will smell fear. If you are stressed, it might make you stress more, and then you could end up releasing chemicals like adrenaline through your glands. Dogs can smell this and know that something is wrong. If you are frustrated and stressed out when training them, they will sense it and it will be harder for them to focus.

Never stop playing with them! Get their energy out through playtime, not through times when they are rough and constantly jumping on others.

Feed them after you eat your own meal. As we already discussed, this is common for the leader of the pack. If you want your dog to really listen to you and remain obedient to what you are asking them, you have to ensure you are establishing dominance over them in a way they respect. You can also take them for a walk right before they eat so that they know you are in charge. You are making them work for their food!

Inconsistencies

The biggest killer for your dog's training is not being consistent throughout the process. Dogs need routines and they only know what boundaries are in place by experiencing them on a repeated basis. If you are consistently inconsistent with your dog and the rules are always changing, they won't know what they should be doing.

What you do at home is what you have to do everywhere else. Even if it might be hard to remember, always keep your dog in training. It will be

Book 3 - Training Your Puppy Step-By-Step

a long process that will take time to get used to. They need to follow certain boundaries in order to stick to the rules which you are teaching them.

Their behavior should be right or wrong at the core. It is wrong to jump up on people. They won't know that it is wrong to jump up on strangers, but okay to jump up on you. They will simply know that it is fine if they do jump.

Even when you don't feel like doing something, you have to stay consistent for your dog. It can be hard to always take them out at the same time, and there are moments where you might just want to stay in bed a little longer even though your dog's already woken you up to go to the bathroom. Even in these moments of struggle, it is essential that you are taking care of your dog's needs and staying stable.

Different partners should not have different rules. This is why it is so important to have a family meeting to establish all of the rules. Even if you have a roommate who's not as interested in caring for the dog, they have to help out just to ensure that no bad habits are created.

Dogs don't have a way to communicate with us other than through signs or symbols. They won't be able to see these if they aren't consistent and presently clear. They need that regulation to understand, as it is really the only true way that they can communicate with you.

Useless Repetition

Your dog needs routine and consistency. However, if you are consistently doing something that's bad for your dog or that they don't understand, then this isn't an ideal routine that will help them in the end.

Useless repetition is like calling their name over and over when they clearly aren't coming. It is like trying to make them sit by pushing down their butts over and over, never able to actually do it on their own. When you notice that you are repeating something to your dog over and over again and they don't get it, then it can be frustrating. However, there are certain steps that you'll want to take instead to get them to really understand what you are trying to teach.

For example, you might find that you are trying to teach them to roll over. You've done it twenty times now and they still don't get it. The first thing that you'll want to do is ensure that you take a break. They might simply be tired, and you could be showing frustration, a bad recipe for pointless training.

Your emotion will change throughout the useless repetition. At first you might be excited with a happy tone while you are training your dog. By the end, your voice might shift and you could end up sounding frustrated. This will only confuse the dog further as they won't really understand what it is that they did wrong to make you frustrated.

Grabbing his paw and shaking it over and over again isn't going to magically work on the 50th time when you are trying to teach him to shake his paw. When something isn't working, you have to reevaluate the process to find a method that they actually do want to participate in.

Sometimes, the repetition is the way that they end up learning. They think that you have to get angry and say 'sit' five times over and over again until they actually decide to sit. Remember that your dog won't think like you do, so they aren't going to understand the pattern you want if you are trying to teach it in an unhelpful way.

Confusing the Dog

It'd be so much easier to train dogs if we knew how to really talk to them. Unfortunately, human-dog communication is still limited and rather than finding methods to speak to them, we have to discover the way that they will understand our signals. Let's discuss what you might be doing wrong which could confuse them. If the dog doesn't understand what you are trying to do, it can make them upset and then it will be harder to train them in the end.

The first thing that confuses them is when you punish them for natural things. This includes things like going to the bathroom, eating, and playing. If you notice that they are going to the bathroom in the house and scream and yell at them, they will think that they're getting in trouble for simply going to the bathroom. If you get home and see that they are eating food you had sitting on the counter, they will think that they are in trouble for eating. You have to catch their bad behavior before it happens, or right as it starts. Punishing them in the middle of it will simply be confusing you.

Did they step on your foot and you yelled? You are not likely to be able to train your dog to not make mistakes. They will still stand behind you when you are cooking sometimes, or they might jump onto your bed and accidentally stick a paw in your gut. Though these accidents are annoying, you can't punish your dog for this because they won't really understand what it is that they're getting in trouble for.

Did they have diarrhea on the floor? They're sick, not intentionally being bad. Though it is bad for them to go inside, they won't be able to hold it all the time, especially if they are ill. The same can be said about them vomiting. Though they might be potty-trained, there are still some accidents that are entirely out of you and your dog's control.

Having the wrong labels for certain cues can make it hard for them to know what to do. Check and ensure that you are not training them to do a certain command with several different words. For example, you might use "sit," "down," or "sit down" all just for the same command for them to sit. Though they all mean the same thing, your dog doesn't know this, and they could end up getting confused.

Try to refrain from using your finger and hand to "punish" them. They will really only understand your use of different words and symbols, not necessarily hand gestures.

Make sure that the affection you are giving them is at the right time. Even though you might want to hold and snuggle them all day long, if they don't seem into it, you can't force it. Your dog might end up thinking that this is some form of punishment, but they won't be sure why you are punishing them in a certain way.

Pay attention to the eye contact that you are giving your dog. They take a lot into consideration when evaluating your eye contact. If you are staring for too long it could make them nervous or anxious. They might even end up getting aggressive depending on the context, thinking that you could be challenging them.

Abandoning a Ritual or Habit

Dogs will need trained throughout their entire puppy-life, and into adulthood. You can't stop training your puppy just because they seem to know the same old tricks. As they get older, you won't have to train with them so often, but there are still important measures to take to ensure they keep up with their knowledge on certain tricks. You might train your dog to play dead within the first few months of their life. If you stop doing this trick, by the time they're two years old, you can't expect them to immediately remember how to do it.

Just because you started once doesn't mean that they will know this forever. In order to ensure that they are keeping up with the tricks that you taught them, have moments where they have to do the treat in order to get a certain reward. The best time to give them trick refreshers is right before they eat, before you take them outside, or after you come back in from taking them out. Give them a treat and positive reinforcement. It only takes them doing these tricks once every couple of days for them to maintain their knowledge of a certain command.

Sometimes, you might abandon an entire ritual. Perhaps you no longer take them outside first thing in the morning and instead wait until it is convenient for you. This can mess up your dog's entire schedule and make them rather confused! If you do need to change protocol, remember to transition your dog. They are already used to a certain way of life and become easily confused. The sudden change could stress them out, so do your best to keep your dog as calm as possible.

Treating All Dogs, the Same

All dogs are different. Even though they have similar needs and will react in usual ways, they're also very different from one another. Some dogs are lazier, some are more aggressive, some are very needy, and others might be independent. Whatever you might have thought about dogs before could be completely wrong after meeting that one special dog that just isn't like any others!

Just because you trained three dogs already doesn't mean this fourth is going to be the same. A mistake that some will make is thinking that all dogs act the same way. They might believe that if a dog isn't acting in the same way that another dog has in the past, that this new dog has something wrong with it. You might have just had a dog that was easier to train first, so the second dog isn't necessarily bad, it is just a little more difficult this time around.

Some dogs will be naturally shyer than others, while some may be boisterous and high energy. Others might simply be ornerier. Some dogs are excited and active, others are scared and want to sit away from everything. Some dogs love new people, some dogs get scared of strangers. Some dogs bark at cars, some dogs like riding in them. Some dogs will play with 100 different toys, some dogs might only want to play with one. Some dogs will be excited to see you, other dogs might not care as much. Each and every dog is different. Not only might they be born that way, but how they were raised before meeting you will also affect their development.

Don't punish your dog because it is not the same as others that you've had. Respect their individualities just as we do for humans! You can assume that some things are going to be similar, especially if that helps in your training. At the same time, remember that some of the things about their personality aren't bad, just different!

Positive Reinforcement with Bad Behavior

One of the biggest mistakes that new trainers can make is that they will end up using positive reinforcement for certain bad behaviors that their dogs have. This occurs when a dog is doing something bad, and the owner gives them a treat in order to stop this bad behavior. In the process, however, the dog learns that if they do this bad thing, they'll get a treat in return.

Sometimes you just want to get your dog to shut up! You just want them to stop barking, maybe they're scratching, or simply being violent. While giving them a treat is a great way to stop them, it also tells them that if they just bark enough, if they get aggressive, or if they start to chew something up, they'll get a reward in return.

Don't use a reward to get them to stop doing something. Instead, use a common command to get them to stop, such as 'sit' or "lay down." Once they have done this good thing, then give them the treat. Next time the dog is barking at cars and you can't get them to stop, don't call them over to give them a treat. Call them over, make them lay down, and then give them the treat. They will better understand that the only way to receive a reward is to be calm and relaxed.

Your dog is smart enough to know how to get what they want. This is just as we mentioned previously with crating the puppy. If they whine and then you let them out, they will know that whining means they can get their way.

Positive reinforcement in negative situations might confuse them as well. For example, let's say that your dog is really afraid of thunder. Whenever there's a storm, they go and hide under a desk. Every time you go to him and say, "it is ok, don't be scared," over and over while you pet him. The petting can help, but they will also pick up on your words. Next time that something scary happens and you tell them, "it is ok, don't be scared," they might be brought back to those same frightened feelings that they had the first time.

Not Proofing Tricks

If you train your dog to sit in the living room within the first three days, that's wonderful! The next step is to 'proof' this behavior. This means that you are taking a trick that they learned and doing it in different settings with various distractions to make sure that they understand the trick at the core. If you teach them a trick in one room, they might only thing that they should be doing that trick in that room. You have to repeatedly do this everywhere so that they understand what the command means at the core.

Dogs don't have the ability to make generalizations like we do. They're not going to understand that just because they learned to 'sit' in the kitchen they're supposed to sit in the living room as well. To really make sure that they're understanding this, you have to proof your tricks.

Start to do the same tricks in different areas throughout your home. Make sure that you do it in the backyard, on walks, and in all rooms.

Add in other distractions after this as well. Do it at the dog park where others are watching. Do it when you have their food prepped and ready to give them in their bowl. Create methods of excitement so that they understand they should be present in all moments and focusing on you.

Do it without a treat in between sessions. You don't want them to stop doing the trick just because you stopped giving them treats. You might have to teach your dog the trick over again, but this is the chance for you to strengthen their abilities. Patience is going to be key! Don't forget to proof or else your hard training work will be lost. If you are giving them treats and showering them with positivity, then they are understanding that this anxiety is fine to have in these scenarios. While it is not a bad thing for them to be scared, don't use positive reinforcement or else they will react the same way next time.

The best thing you can do to comfort your dog is to distract them. Get their favorite toy and play with them. Give them commands and tricks, rewarding that behavior with treats. Act normally and continue to be excited, not sad or scared. Don't force them either. Sometimes it is best to leave them alone while still being in the same room just so that they know you are there to keep them safe. You might put your arm around them or pet them, but refrain from using positive language or treats because they will get the idea this is how they should act in fearful situations, making them only more scared in the end.

Waiting Too Long

If you wait too long to start to train your dog, you are wasting precious time that could be spent helping them grow. Old dogs can learn new tricks, but it is going to be so much harder. Those who wait might grow frustrated with their pups easily and think that it is the dog's fault they aren't learning fast enough.

Start to train them from the moment you move them in. Rather than associating the different tricks with just you, they will associate this new behavior with their home. It is a great fresh start for them, especially when adopted, so take this opportunity to create good habits right away.

Be consistent! You can't get mad at your two-year-old dog that it can't walk on a leash still if you never trained it to properly use a lead.

This is the same for those that wait too long to 'punish' their dog. If you come home to find the garbage has been torn through, they're going to be excited to see you because they probably forgot about this. You might get upset at them right away, and then they will think that you are simply upset to see them.

Though dog-shaming videos and memes are cute, dogs aren't reacting this way because they feel remorse over what they did. The reaction of the sad face and low head are only because they see that you are upset, and they're scared. No matter how guilty they might appear to be, they are really just making that expression because they are afraid of your mood (Manning, 2014).

Chapter 9: Your Step-by-Step Training Plan

Congratulations on making it to this part of the book! You are almost ready to bring your puppy home. The most important part is that you create a schedule. We have a how-to on the way that you can do this, along with a few other key principles to remember before we finish!

Get to Know Your Dog

As we already mentioned, all dogs are different. When you first bring your dog home, make sure that you are doing your best to really get to know them. They will have their own fears, strengths, weaknesses, and interests. Some dogs will be the type that always want to be outside, and others will be perfectly fine with sitting on the couch and sleeping all day. While we did our best to ensure this would be a comprehensive guide for all new pet owners to know how to train their dogs, you should still look into further research to discover the methods that you could use for specific breeds. Though it doesn't matter as much what random facts you might know about your dog, knowing their history, their intended breeding purpose, and so on can really help you better understand why your dog might operate in the way that it does.

Create a Schedule

It is important that you keep your dog on a schedule, just as you would a baby or a toddler. It is not only a good way for you to get into better habits, but it is the method that your dog is going to be able to learn how to function properly as well.

You should start to feed your puppy four times a day until they are around three months old. The best time to do it is when they first wake up and a couple of hours before bed. Then, you can do it twice in between whenever it is convenient for you, so long as it has been spaced out.

Next, until they are six months, drop down to three times a day. The same rules apply when you are figuring out when to feed them. It should be at times that are convenient for you in order to ensure that it is a schedule you can keep up with. For example, if you are usually heavy into work or studying around noon every day, don't try to disrupt your schedule with a feeding here. Though it might be a better time for your pup, if it is not a good time for you it might get forgotten or overlooked, resulting in inconsistencies which confuse your puppy.

Once you get them fixed, you can cut it down to twice a day. Make sure that you adhere to serving suggestions on the packaging of your selected food so as to not overfeed them. eight hours is usually a good time to feed them. After they have eaten, you should take them out around two hours later, though when training puppies, every hour is ideal if you can. They have smaller bladders and aren't going to be able to hold it. As they grow older and learn to control their bladder movements, you can reduce these time periods so that they learn how to resist the urge to go when they have to.

Here are all the things that you will want to include in your schedule for a puppy under six months:

- Three to four feedings
- Five potty breaks
- Training session
- Exercise routine
- Playtime
- Bed time

Again, it will really be dependent on your specific and individual schedule when these times might be. Stick to what works best for you and always consider your dog's needs. Here's an example of what a specific schedule might look like:

- 8 AM: potty break
- 8:30 AM: feed
- 10:30 AM: potty break
- 11 AM: training session
- 1 PM: feed
- 1:30 – 1:45: crate
- 2:00 PM: playtime
- 3 PM: potty break
- 3:30 PM – 3:45: crate
- 4 PM: potty break
- 5 PM: feed
- 7 PM: potty break + 30 minute walk
- 8 PM: feed
- 9 PM: playtime
- 10 PM: potty break
- 10:15 PM: bed time

This can seem like a lot, but feedings and potty breaks will be reduced as your puppy grows. Play around with your schedule at first, but after about a week of being in their new home, your puppy should have a set schedule that they can depend on.

Keep it Simple

Don't overly confuse your dogs. Always look for ways to keep things simple, short, and sweet because this is the method that they are going to be able to learn from the best. If you discover that they are having trouble learning a new trick, it could be that you are making it too complicated. If you find that it is too hard to keep them on a schedule, it might be something too complex for you. Though this book has been comprehensive and long, it is also not as hard as it might seem to keep your dog on a schedule and properly trained. With effort and some troubleshooting, you'll be able to find the perfect routine for you and your furry friend.

Work on Rewards

Remember that positive reinforcement is the center of proper training. When you are able to provide them with rewards they really like, you'll have exceptional results. Experiment with different treats to find the ones they like the most. The more they like a treat, the more likely they are to respond when you are training them! Rewards aren't just in the form of treats either. They're also the positive words and pets that you use for them. Notice the spots that they like scratched or pet the most. Do they prefer their belly rubbed? Their ears massaged? Their chins scratched? When you know your dog's favorite rewards, you can use these to get them to act the way that you want.

Repeat, Repeat, Repeat

Just because your dog successfully does a trick one time doesn't mean that they already know how to do this. They will still need repetition to ensure that they always really understand the trick you are trying to get them to do. though it can seem tedious and at times exhausting, go over and over and over the same trick to make sure that your dog has been able to properly learn it. Like we mentioned earlier, useless repetition will only set them back in their training, so make sure that you aren't repeating something that isn't working.

Check-In with Your Dog

Always consider your dog's needs first and foremost. Get to know their emotions. What do they do when they're anxious? How do they react when they're scared? Are they getting enough attention? Are you showing enough affection? Though it might seem easy to really understand a dog, they are more complex than you'd think! Just because something works for you doesn't mean it does for your dog. When things aren't going right, ensure that they are happy and taken care of. You are their caretaker, so you will always want to do regular check-ups with them to make sure you are providing the best care possible!

Conclusion

Don't feel nervous about getting a puppy, it should be exciting! At the same time, remember that it is crucial you train your puppy in a responsible way and never overlook the important skills they need to be a healthy dog.

Start by bringing home a dog that will work for you in the first place. Though you might like the look of a certain breed, an Australian Shepherd isn't good for someone who doesn't go outside often and enjoys sitting on the couch. Vice versa, if you are someone who enjoys an active lifestyle activities such as exercising, hiking, and so on, a mini poodle that sleeps for 20 hours a day isn't the top choice.

After you've picked your dog, it is time to get your home ready. Pick everything up that could be dangerous, and provide them with a safe space that they don't have to feel uncomfortable in. Give them necessary items and pick out the toys that will help them grow and learn. Give them plenty of treats and water, as well as carefully selected food.

Ensure that you start training from the moment that you get home. Teach them how to sit, and how to go to the bathroom outside right away. Every other trick that you teach them will likely build from sitting down first, and this is a great way that you can make them calm and collected whenever you need them to act that way.

Socialize them from an early age as well. They'll need to get used to controlling their excitement in public, and it is essential that we teach them good behavior with other dogs and people from the very start. This is especially important for aggressive dogs. You have to also ensure you are getting them out of the house to tire them out. Active and ornery dogs may go a bit stir-crazy when you don't get them outside and playing around enough.

Once you have a healthy routine and can start to socialize them properly, it is time to add in more challenging skills. These are things like rolling over or laying down. Once they are able to learn one trick, it will be easier for them to learn even more.

After all of this, ensure that you are giving them the right amount of exercise. Though you might be playing with them and training them at the same time, they will need separate moments of exercise. This could be in the form of agility training or going for basic walks.

At the same time, you'll need to ensure you take them to the vet for regular check-ups and that all of their health needs are being met. Give them proper baths and grooming and don't forget about their need for flea or tick treatments as well.

Never forget the common mistakes that other dog owners might make.

Book 3 - Training Your Puppy Step-By-Step

Each dog is unique and needs to be treated this way, so even though something might work for one dog, that doesn't mean it will for another.

Your dog is going to be your best friend. They will think about you all day long when you are not home. You are the one that they care about more than anything and they put all of their trust in you. Give them the best chance possible at a happy and healthy life by always properly training them.

References

ASPCA. (n.d.). Pet Statistics. Retrieved from https://www.aspca.org/animal-homelessness/shelter-intake-and-surrender/pet-statistics

ASPCA. (n.d.) Vaccinations for Your Pet. Retrieved from https://www.aspca.org/pet-care/general-pet-care/vaccinations-your-pet

Becker, K. (2012). Want a Well-Behaved Dog? Do More of This and Less of That. Retrieved from https://healthypets.mercola.com/sites/healthypets/archive/2012/08/03/positive-reinforcement-dog-training.aspx

Bovsun, M. (2019). How to Potty Train a Puppy: A Comprehensive Guide for Success. Retrieved from **https://www.akc.org/expert-advice/training/how-to-potty-train-a-puppy/**

Dayton, R. (2016). New Study Shows Potential Benefits Of Spaying/Neutering Dogs After Age 1. Retrieved from https://pittsburgh.cbslocal.com/2016/11/14/new-study-shows-potential-benefits-of-spayingneutering-dogs-after-age-1/

Finlay, K. (2017). Can Dogs Taste? Retrieved from https://www.akc.org/expert-advice/lifestyle/can-dogs-taste/

Greenwood, A. (2016). Here Are The Surprising, Preventable Reasons Why People Give Up Their Pets. Retrieved from https://www.huffpost.com/entry/aspca-report-pets-given-up-to-shelters_n_56896846e4b06fa68882a134

Keliher, I. (n.d.). How to Choose the Perfect Dog Name (with Science). Retrieved from **https://www.rover.com/blog/dog-name-advice/**

Manning, S. (2014). Behaviorists: Dogs feel no shame despite the look. Retrieved from https://www.usatoday.com/story/news/nation/2014/02/26/dogs-shame-guilty-look/5833395/

The Humane Society. (n.d.) Crate training 101. Retrieved from https://www.humanesociety.org/resources/crate-training-101

The Humane Society. (n.d.) Fact Sheet: Puppy Mills and Pet Stores. Retrieved from **https://www.humanesociety.org/sites/default/files/docs/pet-stores-puppy-mills-factsheet.pdf**

The Humane Society. (n.d.). Pets by the numbers. Retrieved from **https://www.animalsheltering.org/page/pets-by-the-numbers**

Walsh, K. (2017). Possible Health Isuses in Common Dog Breeds. Retrieved from https://www.healthline.com/health/dog-breeds-and-health-issues

Paul Davis

www.ingramcontent.com/pod-product-compliance
Lightning Source LLC
Chambersburg PA
CBHW072026230526
45466CB00020B/934